The New Mechanical Philosophy

The New Mechanical Philosophy argues for a new image of nature and of science—one that understands both natural and social phenomena to be the product of mechanisms, and that casts the work of science as an effort to discover and understand those mechanisms. Drawing on an expanding literature on mechanisms in physical, life, and social sciences, Stuart Glennan offers an account of the nature of mechanisms and of the models used to represent them. A key quality of mechanisms is that they are particulars—located at different places and times, with no one just like another. The crux of the scientist's challenge is to balance the complexity and particularity of mechanisms with our need for representations of them that are abstract and general.

This volume weaves together metaphysical and methodological questions about mechanisms. Metaphysically, it explores the implications of the mechanistic framework for our understanding of classical philosophical questions about the nature of objects, properties, processes, events, causal relations, natural kinds and laws of nature. Methodologically, the book explores how scientists build models to represent and understand phenomena and the mechanisms responsible for them. Using this account of representation, Glennan offers a scheme for characterizing the enormous diversity of things that scientists call mechanisms, and explores the scope and limits of mechanistic explanation.

Stuart Glennan is the higher Ice Professor of Philosophy at Butler University, and Associate Dean of the College of Liberal Arts and Sciences. Glennan's research has centered on topics in the philosophy of science—especially causation, explanation, modelling and the concept of mechanism. He has also written on science education and on the relation between science and religion.

The New Mechanical Philosophy

Stuart Glennan

OXFORD
UNIVERSITY PRESS

OXFORD

UNIVERSITY PRESS

Great Clarendon Street, Oxford, OX2 6DP,
United Kingdom

Oxford University Press is a department of the University of Oxford.
It furthers the University's objective of excellence in research, scholarship,
and education by publishing worldwide. Oxford is a registered trade mark of
Oxford University Press in the UK and in certain other countries

First Edition published in 2017
First published in paperback 2019

Published in the United States of America by Oxford University Press
198 Madison Avenue, New York, NY 10016, United States of America

British Library Cataloguing in Publication Data
Data available

Library of Congress Cataloging in Publication Data
Data available

ISBN 978-0-19-877971-1 (Hbk.)
ISBN 978-0-19-884807-3 (Pbk.)

For Lesley,
with gratitude for all that you are and all that you do

Contents

Preface

I have been thinking about mechanisms for a long time. I came upon the topic as I was trying to say something about what constituted a causal connection. At that time, outside of the history of philosophy, the concept of mechanism was little discussed. I found my inspirations in various authors—Wesley Salmon, Nancy Cartwright, Herbert Simon, Ronald Giere, and my advisor Bill Wimsatt among others—but a lot of it was just thinking about examples and trying to come up with a coherent theory to account for them. The result was my dissertation (Glennan 1992a). It took some time to find someone willing to publish my first paper on mechanisms (Glennan 1996), and I frankly was not sure anybody was going to read it.

Thankfully, Peter Machamer, Lindley Darden, and Carl Craver did read it. They too were thinking about mechanisms, and they were thorough enough to search the indexes, and find my paper and dissertation. Lindley, who I did not know at the time, called me out of the blue to tell me about their work, and to ask me some questions about my dissertation. Their paper "Thinking about Mechanisms" (Machamer et al. 2000), which was framed in some part as a criticism of things that Bill Bechtel, Bob Richardson, and I had said about mechanisms, linked us together as New Mechanists. Its publication opened the floodgates, and suddenly everyone was thinking about mechanisms.

So it is that twenty-five years later I find myself writing a book about the same topics— mechanisms, models, and causation—that I wrote about in my dissertation. But unlike the first time, I have a very rich literature to draw upon—not just with respect to mechanisms, but also with respect to modeling and causation—two areas of research that have also exploded in the last twenty years. I have benefited immensely from reflecting upon all of this work by proponents and critics of mechanistic approaches— and the result I hope is a far richer understanding of mechanisms and mechanical philosophy than I had when I started.

The idea for this book was born at a conference on mechanisms and causality in the sciences, which was arranged by Phyllis Illari, Federica Russo and Jon Williamson at the University of Kent in September 2009. I remember well a very good pre-conference dinner where one of the other speakers, Stathis Psillos, told me it was time for me to write my mechanisms book, and I agreed. Equally important, a number of the themes in the book are directly traceable to talks and conversations at that conference.

Much of this book was written on a sabbatical I spent at the Konrad Lorenz Institute for Evolution and Cognition Research in 2013. I am grateful for their financial support, which allowed me and my family to spend seven months in Europe, and I owe a great debt to the leadership and staff of the KLI: Gerd Müller, Isabella Sarto-Jackson, Eva Lackner, and especially Werner Callebaut, the then scientific director, who I deeply regret did not live to see this book, which he did so much to foster. The KLI was in its last year in the Lorenz mansion in Altenberg, and the quiet office combined with the daily lunchtime discussions were an ideal place for embarking on this project. I am grateful to all the fellows for their conversations—especially to Tudor Baetu, with whom I shared many a coffee and conversation about mechanisms. Many of the examples I use in this book, both serious and light-hearted, draw their inspiration from Vienna and its environs.

During that sabbatical, I had the opportunity for short stays at a number of universities, where colleagues arranged workshops and reading groups where I could share work in progress and hear about related work. I am particularly grateful to Carl Hoefer and Laura Felline at the UAB in Barcelona, John Dupré at the University of Exeter, Phyllis Illari at University College London, Erik Weber and Bert Leuridan at the University of Ghent, and Marie Kaiser and Andreas Hütteman at the University of Cologne for hosting me. I also received valuable feedback from audiences at talks at the Institute for Philosophy in London, the International Society for the History, Philosophy and Social Studies of Biology, and the German Society for the Philosophy of Science.

I would not have been able to undertake this project without the support of my colleagues at Butler University. My department chairs, Harry van der Linden and now Chad Bauman, have been very supportive, and I received excellent support from my friends in Butler's Institute for Research and Scholarship, as well as Butler's libraries. Special thanks are due to my Dean, Jay Howard. This entire project has been completed during my tenure as Associate Dean of Butler's College of Liberal Arts and Sciences. Jay is one of those rare deans who manages to continue to do research, and he encouraged me to follow the same path, taking the highly unusual step of allowing me to take a sabbatical during my tenure as Associate Dean. I also do not know how I would have survived without the support of the Dean's assistant, Priscilla Cobb, who managed many small miracles, like arranging to replace a dying computer from across the sea, two days before a big talk.

I am very grateful for the comments I received from three readers from Oxford University Press. Carl Craver, Carlos Zednik, and one reader who remained anonymous provided me with very helpful and complementary perspectives on my manuscript. I owe a particular debt to Carl, who, in addition to reading the manuscript for the press, read the manuscript in its entirety on another occasion, and organized two separate visits to Washington University where I could give talks on different portions of the book.

Many people have provided comments on parts or the whole of the book: Marshall Abrams, Ken Aizawa, Holly Andersen, Tudor Baetu, Rafaella Campagner, Lorenzo Cassini, Lindley Darden, Raoul Gervais, John Heil, Phillipe Huneman, Phyllis Illari, Beate Krickel, Meinard Kuhlman, Mark Povich, Federica Russo, Christoph Schultz, Marcel Weber, Rasmus Winther, and doubtless a few others of whom I have lost track. Though flaws no doubt remain, it is an immensely better work for their input.

I am very grateful to Peter Momtchiloff, the philosophy editor at Oxford University Press, UK, who has been very supportive of this project—providing me substantial guidance on both the substance and production of the book. I also want to thank my copy-editor, Kim Richardson, for his careful attention, thoughtful suggestions, and quick responses to my queries. I am also grateful for the assistance of my proofreader, Fiona Tatham, and project manager, Premkumar Kaliamoorthi, who skillfully guided me through the final stages of production.

My greatest debt though is to my family, without whom this book never would have been completed. My wife Lesley and young son Elliot uprooted themselves to join me on the sabbatical where much of the book was written. Elliot braved a German-speaking kindergarten and Lesley managed a four-year-old in a country where she did not know the language. My older sons, Paul and John, who were in college at the time, supported me by taking an interest in my work and by admirably taking care of themselves and each other during a long time apart. In the three years it has taken to complete and revise the manuscript, both Lesley and my sons have been unfailingly supportive of my project—giving me the time and encouragement I needed to make it to the finish line.

1

What Is the New Mechanical Philosophy?

The aim of philosophy, abstractly formulated, is to understand how things in the broadest possible sense of the term hang together in the broadest possible sense of the term.

(Wilfred Sellars 1963)

The past twenty years have seen the emergence of a body of research in philosophy of science that has sometimes been called the New Mechanical Philosophy or New Mechanism. As I shall interpret it, the New Mechanical Philosophy fits squarely within Sellars's conception of the philosophical enterprise. It is an attempt to say how things hang together. The things in this instance are nature and science, understood in the broadest possible senses of these terms. It says of nature that most or all the phenomena found in nature depend on mechanisms—collections of entities whose activities and interactions, suitably organized, are responsible for these phenomena. It says of science that its chief business is the construction of models that describe, predict, and explain these mechanism-dependent phenomena.

My aim in this book is to provide a systematic exposition of the New Mechanical Philosophy—its central motivations, problems, and prospects. While my views owe a great deal to the work of others who have been called New Mechanists—especially Bill Bechtel, Carl Craver, Lindley Darden, Peter Machamer, and Bob Richardson—this remains my account. Along the way, I will try to honor my debts by pointing to areas of consensus and disagreement, but my main goal is to craft a picture of nature and science that is both empirically informed and metaphysically and epistemologically coherent.

What is perhaps most distinctive about my account is its scope. The New Mechanism has developed chiefly as a set of views about explanation and discovery in biology and the life sciences—its primary domains of application have been molecular biology, neuroscience, and cognitive science. Given the ubiquity of mechanism talk in disciplines as diverse as historiography, sociology, and physics, it is natural to wonder whether the insights of the New Mechanism apply more widely, or whether mechanisms and mechanistic explanation are distinctive features of some special sciences. I will argue that there is an informative way of thinking about mechanisms across the sciences—one that will support the philosophical aim of seeing how things hang together. Ironically though, one of the most important general things we can say about mechanisms is that they are particulars—each one different from the next. Sorting out the tension between the particularity of mechanisms and the generality of the ways we represent them will be a central task of this book.

1.1 The Craving for Generality

Writing in the mid 1930s, Ludwig Wittgenstein suggested that one of the chief sources of philosophical error was a weakness he called "our craving for generality" (Wittgenstein 1958, 17). Whether or not it is an error, judging by Sellars's remark from thirty years later, he is certainly right about the craving. Wittgenstein thought of this affliction as arising out of certain bad philosophical habits. For instance, he saw it in "the tendency to look for something in common with all the entities which we commonly subsume under a general term" (17). Relatedly, he saw it in the "tendency rooted in our usual forms of speech to think that the man who has learnt to understand a general term, say, the term 'leaf,' has thereby come to possess a kind of general picture of a leaf as opposed to pictures of particular leaves" (18). Wittgenstein's initial philosophical target is no doubt Plato and his theory of the forms. Socrates's first move, familiar to any reader of Plato's dialogs, was to ask of any concept X (piety, virtue, justice, love . . .) what form all instances of X partook in that made them X. But the craving is by no means limited to Plato, and can be seen in any attempt to offer a general analysis of the nature of Xs, from Descartes's account of the nature of body to Wittgenstein's own earlier attempt to characterize the general form of the proposition. Certainly, I am guilty of it in trying to offer a general account of mechanism.

While part of the source of our craving comes from our philosophical tradition and our forms of speech, it is particularly interesting for our account that he sees the other main source as "our preoccupation with the method of science" (18). He describes scientific method as

the method of reducing the explanation of natural phenomena to the smallest possible number of primitive natural laws; and, in mathematics, of unifying the treatment of different topics by using a generalization.   (18)

He goes on to say:

Philosophers constantly see the method of science before their eyes, and are irresistibly tempted to ask and answer questions in the way science does. This tendency is the real source of metaphysics, and leads the philosopher into complete darkness.   (18)

There is both insight and irony this passage. Scientists do seek generality, and many philosophers, both in Wittgenstein's time and today, deliberately seek to emulate the style and approach of scientists. The irony is in the presupposition of this passage—that there is such a thing as *the* method of science. To assume that the activities of all scientists are instances of some general kind of activity called "the method of science" is once again to commit Plato's mistake. Wittgenstein may be forgiven, as he was writing during the heyday of logical positivism and the unity of science movement—a movement that above all things exemplified the craving for generality. Subsequent research in the philosophy and history of science have taught us that there is not one scientific method but many, though these diverse methods may share, to use Wittgenstein's apt phrase, rich family resemblances.

The craving for generality is more widespread than Wittgenstein suggests. It afflicts scientists as well as philosophers, and not just them. Really the craving for generality is an essential part of the human condition. (The irony of appealing in this context to "an essential part the human condition" is duly noted.) In science, the craving for generality expresses itself in the search for laws and generalizations; in human life generally it expresses itself in the search for patterns. Here is the nub of the problem: reality is particular, but the needs of anyone who seeks to understand and survive in the world are general. If I am to live with tigers, I need to know how to treat them. One tiger is different than another, and the same tiger will have very different properties on different occasions depending upon whether it is young, old, horny, hungry, or grumpy. Still I need general rules like "don't pet tigers" or I am not going to be around for very long.

I shall argue that the insight that reality is particular is at the very heart of the New Mechanical Philosophy. The phenomena that constitute our world are the products of mechanisms: car engines are mechanisms for rotating drive shafts; eyes are mechanisms for transducing light into neural impulses; oxidation is a mechanism that produces rust. And the crucial thing about mechanisms is that they are particular and local. One car (or eye, or oxidation reaction) is not exactly like another. And while we can and must look for similarities that allow us to understand cars or eyes or chemical reactions as types, each instance of these things is a little different, and the source of their causal powers lies in those particular instances. In contrast to Wittgenstein's view, the New Mechanists believe scientific methods are chiefly directed toward the discovery and representation not of laws but of mechanisms. The generalizations we sometimes call laws are heuristic; they do not reflect the deep reality of things.

To see both the sources of our craving for generality and the troubles it brings, consider an example somewhat removed from the natural sciences—the connection between terrorism and Islam. According to a report of the National Consortium for the Study of Terrorism and Responses to Terrorism (START), a research center sponsored by the United States government, there were more than 16,800 terrorist attacks in 2014, which caused more than 43,500 deaths (Miller 2015).[1] A large number of these

---

[1] Such claims have precision only to the extent that there are clear and non-arbitrary rules for counting acts of terror and distinguishing them from other kinds of acts. I shall not say anything about this here, beyond noting the obvious fact that any classification of a set of events as being of a kind like "terrorist acts" involves categorizing heterogeneous particulars according to criteria that are pragmatically driven and value-laden.

attacks are attributable to the activities of Islamic extremist groups, with the top perpetrators being the Islamic State of Iraq and the Levant (ISIL), the Taliban, and Al-Shabab in Somalia. While there are certainly non-Islamic groups that sponsor terrorist attacks—e.g. the FARC in Colombia and ETA in Spain—acts of terrorism in the last decade have been committed disproportionally by people who identify themselves as Muslims.

Despite this data, it is clearly misguided to infer that Islam causes terrorism. For one thing, terrorist attacks are, for all of the fear they elicit, rare events. There are over 1.5 billion Muslims in the world, and even if every one of the 16,800 terrorist attacks in 2014 were carried out by Muslims, the proportion of Muslims engaged in terrorism would be vanishingly small. But the other reason, more important for our purposes, is that there simply is no property of being Muslim which all or most people who identify as Muslims share, and hence it simply makes no sense to say that Islam, as such, either inclines or disinclines its adherents to terrorism.

Like any religious tradition, Islam is an amalgam of beliefs, practices, and identities that share common historical antecedents but which find very different expressions as they have evolved over time and space in interaction with local social, economic, political, and cultural contexts. While it is clear that many fighters acting on behalf of ISIL and other such terrorist organizations are motivated by their understanding of their religious identity, that identity is shaped not by Islam as such, but by the very particular circumstances in which they acquired their identity. If any sort of Islam is causally connected to terrorism, it will be an Islam of a particular place and time, an Islam nurtured in Cairo or Riyadh, or Molenbeek, in which jihadist identities are constructed in response to their particular social, economic, and political conditions.

Racial and ethnic profiling is a psychologically unsurprising but often epistemically and morally problematic expression of our craving for generality. But not all attempts to make generalizations about heterogeneous groups of individuals are either morally or epistemically inappropriate. Border security agents may be wise to concentrate their efforts on people from certain countries or of certain genders or ages, as members of certain groups are indeed statistically more likely to be engaged in illegal or terrorist activities. General if fallible heuristics can, properly applied, increase their effectiveness at detecting real threats. And in a more prosaic context, we profile in all sorts of ways in order to get about our daily lives. For instance, as a teacher I have generalized expectations about the capacities, interests, and aspirations and mores of my students that are based on general categories to which they belong—philosophy majors or biology majors, first years or fourth years, New Yorkers or Midwesterners, and so on. Such generalized expectations are essential for planning courses, though they are sometimes flat-out wrong.

In the context of psychosocial questions like this, the point I am making is uncontroversial: What causes an individual to become a terrorist or a priest or a wine-maker is the product of the very particular history of that individual. We may be able to make statistical generalizations about the relationships between these properties, but such generalizations are simply local summaries of correlations among populations of heterogeneous individuals. The generalizations may be of local use, but they will not be projectable to different times or places. And these generalizations, while they may fallibly predict, will not explain. If we are really to understand what causes an individual to do anything, we must look at the particular properties and circumstances of that individual.

But if reality as a whole is particular, then these features of psychosocial properties and causation are not unique. Just as there is no one thing that it is to be a Muslim, there is no one thing that it is to be a cell or to be an oxidation reaction. Even for physical, chemical, biological, and other phenomena studied by the natural sciences, the causes of these phenomena are local, heterogeneous, and particular. We live, as Nancy Cartwright says, in a dappled world.

1.2 Antecedents of the New Mechanism

The set of philosophical approaches and conversations that have come to be called the New Mechanism originate in the 1990s, and are commonly traced to three publications. Bechtel and Richardson's book, *Discovering Complexity* (1993), provided an account of biological discovery that characterized much work in biology as a search for mechanisms. Around the same time, I argued that causal relations are best understood as depending upon mechanisms (Glennan 1996). Finally, in 2000, Machamer, Darden, and Craver, often referred to collectively as "MDC," published their immensely influential paper "Thinking about Mechanisms" (Machamer, Darden, and Craver 2000), which offered itself as a kind of manifesto for a new approach to philosophy of science.[2]

As with any other "ism," there is both value and danger in pulling together a set of not always consistent strands of thought under a name like "New Mechanism." Clearly the philosophers who have been called New Mechanists do not have identical views or interests, and their own views have in fact changed over time. In this book, as I articulate my own account of the New Mechanical Philosophy, many of these points of difference will emerge. At the same time, I think it is fair to say that the New Mechanists—MDC, Bechtel and Richardson, myself, and others—have been part of a common conversation and share many common commitments that distinguish this recent brand of Mechanism from other mechanical philosophies and other styles of philosophy of science. To situate this conversation, let me say just a bit about these common commitments and their relationship to other mechanical philosophies and other recent developments in the philosophy of science.

---

[2] To my knowledge, the term "New Mechanistic Philosophy" ("Mechanistic," not "Mechanical") first occurs in Skipper and Millstein (2005). Jim Bogen used the term "New Mechanical Philosophy" as a title for his review of a collection of essays by Darden (Bogen 2008b). In other literature, members of this group are referred to as "Mechanists," e.g., Psillos (2004) or, tongue in cheek, as "Mechanistas." For brief overviews of the New Mechanism, see Craver and Tabery (2015); Glennan (2015b).

Mechanical philosophies and mechanistic approaches to understanding nature begin in antiquity with the atomism of Democritus and the Epicureans (Popa 2017). The heyday of mechanical philosophy was in the seventeenth century, espoused in various ways by, among others, Galileo, Thomas Hobbes, René Descartes, Pierre Gassendi, and Robert Boyle. Not without reason, many historians of science see "the mechanization of the world picture" as a central aspect of the birth of modern philosophy and of the scientific revolution (Dijksterhuis 1961; Westfall 1971; Shapin 1996; Roux 2017). Many but not all seventeenth-century versions of mechanical philosophy are closely connected with atomic or corpuscular theory, holding that all visible natural phenomena are the product of the motions and interactions of microscopic atoms or corpuscles of various sizes and shapes.

While the seventeenth-century mechanical philosophy was chiefly concerned with explaining manifest physical and chemical properties of inanimate objects (heat, light, color, gravity, etc.), discussions in the eighteenth and nineteenth centuries increasingly focused on mechanical philosophy as a way to understand the nature of living things. These mechanists emphasized the continuity between living and non-living things. They shared with the seventeenth-century mechanists a suspicion of final causes, and they sought to explain the behavior of living systems by analogy to physical systems and especially to machines. A classic polemic in this vein was the French physician and philosopher Julien Offray de La Mettrie's 1747 treatise *L'Homme machine*. Mechanism in this sense is opposed by vitalism or organicism, views that suggest that there is something essential (goal-directedness, self-organization, etc.) that renders living things of a fundamentally different kind than physical systems.[3]

While there are certainly historical and conceptual connections between the New Mechanism and these earlier incarnations of mechanical philosophy, there are important differences that bear emphasis at the outset. First, the New Mechanists are not committed to atomism either metaphysically or methodologically. New Mechanists have emphasized that nature is hierarchically arranged, with new and different kinds of entities and interaction arising at different levels of organization. Whereas many seventeenth-century mechanists were committed to the idea that all phenomena could in principle be explained in terms of action by contact of variously shaped microscopic corpuscles, the New Mechanists think of mechanisms involving objects of diverse kinds and sizes (molecules, magnets, cells, organisms, stars) engaging in a variety of different kinds of activities and interactions (chemical bonding, electrical conduction, absorption, coagulation, predation). While New Mechanists believe that these objects and their activities and interactions are composed of and explained by the activities and interactions of their constituents, they are not committed to atomism. Second, the New Mechanists emphasize that there are important differences

---

[3] For helpful discussions and further references on the historical development of mechanistic thinking in biology see Craver and Darden (2005); Nicholson (2012); Allen (2018). For an illuminating history of the concept of self-organization and its relation to mechanisms in the physical sciences, biology, and engineering, see Keller (2008; 2009).

between mechanisms and machines. Human-built machines have mechanisms by which they operate, and the machine metaphor can be helpful in understanding the behavior of many naturally occurring mechanisms; however, there are many kinds of mechanisms both in living and in non-living systems that do not behave like the kinds of mechanisms we find in windmills, cars, or toasters.[4]

The New Mechanism is not the only outbreak of mechanical philosophy in recent decades, and several related approaches should be mentioned. Salmon saw his approach to causation and causal explanation, developed in his seminal *Scientific Explanation and the Causal Structure of the World* (1984), as a reincarnation of mechanical philosophy, and the terms "mechanical philosophy" and "the causal-mechanical account of explanation" have often been used to refer to this work (Hitchcock 1995; Woodward 1989). Although principally concerned with explanation in the physical sciences, Salmon's approach shares many features with the New Mechanist approach.[5] Two other strands of mechanical philosophy are mechanistic approaches to cognitive science and medicine (Thagard 1999; Thagard and Kroon 2006; Russo and Williamson 2007), and a burgeoning literature on mechanisms and mechanical explanation in the social sciences (Hedström and Ylikoski 2010; Little 2011; Ylikoski 2017).

Although the explosion of interest in mechanisms has occurred in just the last twenty years, the New Mechanism and other recent mechanistic approaches reflect a continuation of trends in philosophical thinking about nature and science that has occurred since the 1960s. I will summarize those trends as a shift in focus amongst philosophers from thinking about laws to thinking about mechanisms and from thinking about theories to thinking about models.

The laws and theories image of science, which finds its most mature expression in late works of logical empiricism, such as Hempel's *Aspects of Scientific Explanation* (1965) and *Philosophy of Natural Science* (1966) and Nagel's *The Structure of Science* (1979), goes something like this: The sciences are ultimately concerned with the observation, explanation, and prediction of phenomena in the natural world. Observations are expressible in terms of singular statements, ideally representable in the language of predicate logic. To understand and explain these phenomena, we must generalize. We start with empirical generalization (or experimental laws) where we identify patterns in observable phenomena. But to extend our scope of explanation and prediction, we must develop theory. Theories are understood to be collections of laws (theoretical

---

[4] It is still a point of controversy whether the New Mechanist approach is really adequate to explaining the character of living systems. While critics do not doubt that there are mechanisms at work in living systems, some believe that the New Mechanist approach is ineluctably entangled with the machine metaphor and that some of the characteristics of living things as wholes (especially their capacities for self-organization, self-maintenance, and self-direction) are beyond the scope of mechanistic explanation (Dupré 2008; 2013; Nicholson 2012; 2013). For one explication and defense of a New Mechanist approach to biological systems that answers some of these challenges, see Bechtel (2011).

[5] Craver in particular (2007; 2013a) sees his approach to mechanistic explanation as building on Salmon's approach. On the relation of New Mechanist approaches to Salmon see Campaner (2013); Glennan (2002a). I address the relation between Salmon's and my views of causation in Chapter 7, and explanation in Chapter 8.

generalizations) concerning entities not directly observable (like gravity or genes). But these theories, in combination with bridge principles connecting theoretical to observable terms, allow us to predict and explain—unifying disparate phenomena under a single structure.

The mechanisms and models approach is distinguished from the laws and theory approach chiefly by its emphasis on particularity. Mechanisms are situated at particular locations in space and time, and while we may classify mechanisms into kinds, ultimately the properties of individual mechanisms, including their causal powers, are heterogeneous and local. There are few if any true laws. Generalizations may have heuristic value, but they don't represent the deep reality of things. The move from laws to mechanisms is paralleled by the move from theories to models. Models (as they are most often understood) are models of particular mechanisms—the Newtonian model of the solar system or the plate tectonics model of continental drift. And while it is routine for scientists to construct generalized models—e.g., models of a type of cell or a type of star—it is understood that modeling involves approximation and idealization.

Many philosophers have contributed to the shift to the mechanisms and models approach. A few of the more important motifs and representative references are the following:

- The world is complex and often disordered, and cannot be characterized in terms of global patterns or regularities; the regularities we find are of limited scope, and our characterization of them reflects our practical concerns (Cartwright 1983; 1999; Dupré 1993; Wimsatt 2007; Mitchell 2009).
- The special sciences are autonomous disciplines with unique problems and methods; there is no one scientific method (Putnam 1973; Fodor 1974; Kitcher 1984).
- There are few laws in the special sciences, and those that can be found are effects to be explained rather than deep truths that explain (Giere 1995; Cummins 2000; Wimsatt 2007).
- Inter-level reduction is best understood not as a relation between theories but as a relation between phenomena and the mechanisms that produce them; reductionist research strategies are heuristic and do not undermine the reality or autonomy of higher-level phenomena (Wimsatt 1972; 1980).
- Science seeks to understand the causes of things, and causes cannot be reduced to regularities (Cartwright 1983; Salmon 1984; Dowe 2000; Woodward 2003).
- Models that are idealized, heuristic, and often clearly false are key tools for representing and intervening in nature. A given system will typically have not one model but many (Cartwright 1999; Morgan and Morrison 1999; Giere 2004; Weisberg 2007a; 2013; Wimsatt 2007).

In claiming these ideas as part of the intellectual provenance of the New Mechanism, I should hasten to emphasize that some of these authors have been explicitly critical of some claims of the New Mechanists, and others have research agendas that take them in different directions. Still, their contributions form the intellectual milieu of the New Mechanism.

1.3 The Philosophy of Nature, the Philosophy of Science, and Naturalism

As I understand it, the New Mechanical Philosophy is a view both about nature and about science. The philosophical views in this book are accordingly at the intersection of natural philosophy and the philosophy of science. Let me say something about what I take these fields to be and the methods I will use to approach them in this book. I use the term "philosophy of nature," and the related term "natural philosophy," to refer to philosophical inquiry into questions about the constitution of things in the natural world—i.e., the world that is the object of scientific investigation. That world is the whole world; it encompasses not just physical nature, but also the world of living things, of human beings and their societies, and the artifacts of human culture.

The term "natural philosophy" was popular at a time when there was no separation between philosophy and science. The early mechanists—Galileo, Hobbes, Descartes, Boyle, Newton—would all count as natural philosophers. Despite their differences, all of these philosophers espoused a natural philosophy that was concerned with characterizing how things in nature worked, and all of them saw continuity between what we would think of as philosophical or metaphysical work and empirical science. While seldom used to characterize contemporary philosophy, the term "natural philosophy" seems appropriate in light of what Werner Callebaut (1993) has characterized as "the naturalistic turn" in philosophy. Naturalism, in the sense that Callebaut describes, is primarily a methodological approach. It denies that there is a clear division of labor between philosophy and science. There is no first philosophy. Conceptual analysis and ordinary language may tell you what people mean, but they do not necessarily tell you how things are. The sciences collectively provide us our best source of knowledge about what things there are in the world and how they work, and metaphysical speculations that do not attend to that knowledge are groundless. On this view, natural philosophy simply represents the theoretical and conceptual pole of the philosophical-scientific enterprise.[6]

Although many philosophers of science engage in what I am calling natural philosophy, I want to distinguish natural philosophy from philosophy of science by their different objects of concern. Natural philosophy is concerned with nature, while philosophy of science is concerned with science. Of course, from a naturalistic point of view, science is a part of nature because the human world is part of the natural world;

---

[6] This conception of natural philosophy is in keeping with the pursuit of "naturalistic metaphysics" or "the metaphysics of science." These are good things, as opposed to "scholastic metaphysics," which is bad. For two opinionated polemics against scholasticism, one from philosophy of physics and one from philosophy of biology, see Ladyman and Ross (2007); Callebaut (2013).

science and its institutions are products of human culture, expressions of the extended human phenotype. Nonetheless, it is very helpful to separate out the peculiar part of nature that is science, so that we can understand its activities and products. To put the matter rather generally, science is a set of institutions created by human beings and their culture to help represent, understand, predict, and control the world. In keeping with the naturalistic denial of the division between philosophy and science, we can profitably think about the philosophy of science as part of "the science of science"— that is, the naturalistic and largely empirical study of science as a natural phenomenon. As such, the philosophy of science is the theoretical wing of science studies, but shares the same building with its other disciplines, like the history, sociology, and anthropology of science.

The account of nature and of science that I will offer presupposes a minimal form of scientific realism. It supposes that mechanisms and their constituents are things in the world that exist independently of the models we make of them. I will, to use a phrase of Stathis Psillos, adopt a realist framework. Psillos characterizes it as follows:

> The realist framework...is the framework that posits entities as constituents of the commonsensical entities and relies on them and their properties for the explanation and prediction of the laws and the properties of commonsensical entities. Accordingly, the realist framework is an explanatory framework, viz., a framework of explanatory posits. In particular, it is a framework that explains by positing constituents of macroscopic things.   (Psillos 2011, 303)

Psillos's view is that this framework is indispensable to this very common sort of explanatory approach within science. I concur, and would add that the explanatory goals for which this realist framework is indispensable are central to the New Mechanist's conception of the scientific enterprise. The realist framework is implicit in the assertion that mechanisms are things in the world that are actually responsible for the vast variety of natural phenomena.[7] But to adopt such a framework does not imply that the entities posited by any particular model or theory refer. Neither is it to suppose that our ways of representing nature are not deeply affected by our interests, cognitive capacities, and other factors. To carefully explore the relationship between a mind-independent reality and our inherently perspectival representations of that reality will be a central task of this book.

For our investigation to be successful it will be crucial to disentangle three kinds of questions about our objects of study: semantic questions, ontological questions, and epistemological questions. Semantic questions are questions about concepts and meaning: what do we (philosophers, scientists of various persuasions, the folk) mean by the term "mechanism" ("cause," "process," "activity," etc.)? These questions are largely distinct from ontological questions—questions about what mechanisms (causes, processes, activities, etc.) are as things in the world. And these questions are in

---

[7] For an instrumentalist approach to mechanisms, see Colombo, Hartmann, and Iersel (2015).

turn distinct from epistemological questions about how we come to believe and know things about these objects.

My primary focus will be on ontological questions. The first question in this book is what a mechanism is as a thing in the world, and this will spur other ontological questions: What is it to be a part of a mechanism, a causal interaction, a system, a cause, and so forth. But these questions, while distinct, are not and cannot be isolated from the semantic and epistemological questions. To develop a coherent and plausible ontological account requires a proper understanding of how these questions bear on each other.

A view commonly held amongst analytic metaphysicians is that the proper starting point for ontological investigation is conceptual analysis; we get to the ontological questions by way of the semantic questions. David Lewis's views on causation provide a good example. Lewis (e.g., 1973; 1979) offers a reductive analysis of causation in terms of patterns of counterfactual dependence. It is reductive in the sense that singular causal claims (like the claim that Caesar's crossing the Rubicon caused the conspiracy to form) can be translated into a set of counterfactual claims (e.g., that had Caesar not crossed the Rubicon, the conspiracy would not have formed). By further analysis we find the truth conditions for counterfactual claims—and it is these that show what our causal claims are really about.

A version of this approach that has received considerable discussion in the last decade is often known as "the Canberra plan." The idea of Canberra planning is to provide a division of labor between the philosophers and the scientists. Philosophers go first—setting the stage by offering an a priori analysis of our concepts. This analysis explicates the role that a certain concept plays within our broader conceptual economy. Then we can turn the question over to the scientists (or to empirically informed philosophers perhaps) who can go look for what it is in the world that fills the role. A classic example of the method comes from Lewis (1972), who characterizes what it is to be a pain analytically in terms of its functional role (pain is what is caused by this and that, and causes this and that other thing); he then leaves it to the scientists to find what it is in the brain that plays this functional role.

In the recent literature on causation we find another notable outbreak of Canberra planning, particularly in the work of Peter Menzies (1996; 2009). Menzies argues that our starting point for understanding causation should be a folk conception, discernible from what he calls platitudes—for instance, that causes precede and are distinct from their effects, or that causes make their effects more likely. These platitudes provide a description of the functional role of the causal relation—an account of what it does. Given this conceptual analysis, we then can look to science—in the form of what Phil Dowe (2000) has called empirical analysis—to find out what sort of features of the world could fit the bill.

There are definitely some things that are right about the Canberra approach. In the first place, it illuminates the difference between ontological and semantic questions. To cite a familiar example, it helps us see that water can be H_2O, even if being H_2O is not, or at least was not until the chemical revolution, part of our concept of water. Additionally, it gives us an account of how our concepts (even our folk concepts) help fix the referents of claims involving those concepts.

The trouble with Canberra planning is that the division of labor between the philosophers and the scientists, and between conceptual analysis and empirical investigation, is not as clean as examples like these suggest. There is for one thing the fact that our folk intuitions may be ambiguous and inconsistent; they may be in need of what Carnap famously called "explication." We also have the worry that not all concepts we employ need refer. We might have concepts like "witch" or "soul," but having these concept does not imply that there are any witches or souls to answer to them. And then there are the intermediate cases. If we have, for instance, a folk concept of belief, and our scientific investigations of animal behavior suggest that there is not anything in our mind-brains that quite fits the bill, what are we to infer—that really there are no such things as beliefs, or just that our beliefs about beliefs were wrongheaded?

Concepts, especially as they are employed in the sciences, have a way of moving around. Take any interesting concept found within the sciences—e.g., electricity, force, compound, gene, species, sensation, memory—and you will find that these concepts have histories, histories that are intertwined with empirical research in the fields. As Quine argued many years ago (1951), there is no clean break between the analytic and the synthetic.

Where does this leave us? I suggest the following: Conceptual analysis has an important role in fixing the topic of our investigation, but this analysis is not the special province of philosophers, and it gives us no immediate insights into metaphysical questions, like what mechanisms or causes are as things in the world. We should begin our ontological investigations with the assumption that the well-confirmed claims of science and common sense are true, and we seek to understand what would have to be the case for these claims to be true. As we come to understand these conditions we will sometimes find that the things in the world do not quite line up with our concepts— and if so our concepts will be in need of revision. This is the back and forth that I understand to be characteristic of natural philosophy.

In focusing on ontological (and more broadly metaphysical) questions, I am departing from an emphasis found in much of the other New Mechanist literature.[8] The New Mechanism emerged from a focus on scientific practice, and has been able to make considerable progress by attending to the methods of discovery, reasoning, and representation in specific cases of mechanistic science. But as this research has progressed, I think it has been clear to many participants in the discussion that metaphysical questions are unavoidable. In writing this book, I hope to show both how examinations of

---

[8] What counts as ontology and metaphysics is by no means settled—and depending upon your views about metaphysics, these fields may be more or less coincident (Chalmers, Manley, and Wasserman 2009). As I understand it, ontology is concerned with what there is, while metaphysics is somewhat broader in considering other questions—like questions about relations of identity and dependence.

scientific theories and practice provide good reasons for adopting certain metaphysical positions, and how clarifying these positions can help illuminate the meaning of those theories and practices. As such, I hope this book will be of interest both to philosophers of science who have seen the need to get clearer on these metaphysical questions, and to metaphysicians who are interested in what mechanistic philosophy of science can do to illuminate their questions. And while I will sometimes dig deep into debates within philosophy of science, I have tried to write this book with language and examples that will make it broadly readable—in part in the hope that philosophically minded scientists will find here an interesting philosophical conception of the scientific enterprise.

1.4 The Book in Outline

This distinction between things in the world and scientific representations of those things is crucial to the argument and exposition of this book. Mechanisms in the world are one thing—a matter of natural philosophy—and scientific representations of those mechanisms, our models, are another. Our first task is ontological—to explore mechanisms in the world, and to understand the relationship between mechanisms and causation. Our second task is epistemological and pragmatic—to understand the means by which scientists construct representations of mechanisms and mechanism-dependent phenomena, and use them to understand, predict, and control things in the world. But while we must distinguish conceptually between these two tasks, we cannot tackle them in isolation, since our ontological account should both make sense of and constrain our epistemological practices.

Here, more specifically is how we will proceed:

Chapter 2: Mechanisms

Chapter 2 provides an account of what mechanisms are as things in the world. It begins with a characterization that I call minimal mechanism, which says that a mechanism for a phenomenon consists of entities (or parts) whose activities and interactions are organized in such a way that they are responsible for the phenomenon. This characterization is minimal both in the sense that it represents broadly shared commitments among New Mechanists as to necessary (but not necessarily sufficient) features of a mechanism, and in the sense that it is permissive, allowing essentially all causal processes to count as mechanisms. The bulk of the chapter is devoted to clarifying this characterization by an explication of the concepts appealed to—phenomena, entities, activities and interactions, and organization. A central theme in this discussion will be the hierarchical character of mechanistic organization—the idea that the entities, activities, and interactions of which mechanisms are made are themselves constituted by mechanisms. The upshot is a "new mechanical ontology"—a view of what there is that is sometimes at odds with traditional ontological presumptions of metaphysics and the philosophy of science.

Chapter 3: Models, Mechanisms, and How Explanations

This chapter shifts the focus from an ontological concern about the nature of mechanisms in the world to a methodologically and pragmatically focused discussion of how scientists represent mechanisms via models. I offer an account of models based upon Ron Giere's pioneering work (Giere 1988; 2004), using it to distinguish between mechanistic and non-mechanistic models. This account of modeling suggests some revisions to claims of New Mechanists about what is required for a model to be mechanistic. On some readings, to give a mechanistic model is to fill in all the details of the entities, activities, and interactions that are responsible for a phenomenon. I argue that for many phenomena that scientists study, it is misleading to take such a goal as a normative ideal. Information about complex and multi-level mechanisms often requires multiple models, and the desire for models that are tractable and general may make abstract and idealized models preferable to more detailed ones. Mechanistic models are the vehicles for conveying mechanistic explanations, which are distinguished from other forms of explanation by the fact that they show how the phenomenon to be explained comes about.

Chapter 4: Mechanisms, Models, and Kinds

Mechanisms, as well as the parts, activities, and interactions that make them up, are particulars; and yet, it is part and parcel of scientific practice to cluster these particulars into kinds. In this chapter I explore the principles that underlie these classifications. I advocate a "models first" approach to mechanism kinds, where particular mechanisms fall under a kind in virtue of their being adequately represented by a certain kind of model. Mechanism kinds have no independent existence apart from their instances (they are not universals) and particular mechanisms are instances of a variety of different and overlapping kinds. Still, mechanism kinds are real in the minimal sense that they are grounded in the objective similarities that exist between particulars that allow them to be represented by common models. In addition to discussing mechanism kinds, I also discuss the way in which the entities and activities that are constitutive of mechanisms may be classified into kinds. As with mechanisms themselves, activity and entity kinds are abstractions, but abstractions grounded in objective similarities between fully concrete particular entities and activities.

Chapter 5: Types of Mechanisms

The strategy of minimal mechanism, as described in Chapter 2, is to offer a permissive definition of mechanism that encompasses a wide and diverse range of systems and processes. While this strategy has the virtue of honoring the diversity of mechanism talk in the natural and social sciences and providing an account suitable for framing a mechanistic approach to causation, there is a danger that the account is so abstract as to be uninformative about the features of nature and science that it seeks to illuminate. In this chapter, I begin to address this problem by offering an account of types of mechanisms. I will suggest that mechanisms can be classified into types along several taxonomic dimensions—the types of phenomena produced, the types of entities and activities of which the mechanism is composed, the types of mechanistic organization, and the types of etiology, i.e., ways in which those mechanisms came to be. I will, following the models-first strategy, show how particular mechanisms can be classified into kinds via similarities in the models that are used to represent them. This account will provide us insights into the various ways in which the sciences are unified and disunified.

Chapter 6: Mechanisms and Causation

This chapter will sketch a mechanistic account of causation and situate it within the space of approaches to causation found in the contemporary philosophical literature. The account suggests that mechanisms provide the truth-makers for a variety of kinds of causal claims, and in particular that claims that one event caused another are made true by the existence of a mechanism by which the first event contributes to the production of the second. The particular and productive character of mechanisms in fact implies that we should think of causation as fundamentally a singular and intrinsic relation between events, rather than as something mediated by laws or universals. The virtues of the mechanistic approach will be illustrated by means of a comparison with counterfactual and regularity approaches to causation. The tension between singularity and generality is again shown in the contrast between the generality of evidence and the singularity of mechanisms and causal relations.

Chapter 7: Production and Relevance

It is now increasingly common to hold that there are two distinct concepts of cause, which I refer to as production and relevance. In this chapter I will explore how the mechanistic approach makes sense of these concepts and the relationship between them. On the mechanistic approach, production is the primary causal relationship, so much of the chapter will be devoted to developing a mechanistic account of production: How and in what sense does the productive character of a mechanistic process arise from the productive capacities of its parts and sub-processes? How can one appeal to the productivity of parts to explain the productivity of mechanisms as wholes without endorsing an overly reductionist thesis that "real causation" is all in microphysics? And, if mechanistic productivity is accounted for by the productive character of parts, what sense can be made of productivity if and when one runs out of mechanistic decompositions? The chapter will also discuss how a mechanistic approach to causation can make sense of intuitions about causal relevance. I will describe and assess standard objections to productive accounts of the nature of causation, arguing that the mechanistic account of production, in contradistinction to theories of physical production (e.g., Dowe 2000), is able to avoid problems of explanatory irrelevance and handle problems like causation by omission and disconnection.

Chapter 8: Explanation: Mechanistic and Otherwise

In the final chapter of the book, I will explore how scientific explanation works in a world made of mechanisms. I shall argue that explanation, in its most general sense, is an activity that involves constructing models that represent dependence relations. Explanation is a matter of showing what depends upon what. I use this basic conception of explanation to address two topics that have been much discussed in the recent literature. First, I attempt to recast the debate over the so-called ontic and epistemic conceptions of scientific explanation, and to revive Salmon's seldom-discussed modal conception of explanation. Second, I show how even in a world full of mechanism-dependent phenomena, many explanations will be neither causal nor mechanistic.

1.5   Conclusion

Our quick tour of the plan of this book may have already suggested a kind of paradox in its aims. While I have argued that recognizing the variability of mechanisms is central to overcoming our craving for generality, I offer the New Mechanical Philosophy as a highly general account of nature and of science. Even to a casual observer, it should be clear that such an effort goes against prevailing trends towards specialization both in the sciences and in naturalistic approaches to the philosophy of science. It also runs counter to much of the work of my fellow New Mechanists, who have tended to offer accounts of mechanisms, mechanistic explanation, strategies for mechanistic discovery, and so on, that are grounded in particular scientific disciplines or sub-disciplines.

In my defense, I can only repeat that there is nothing irrational in the craving for generality, so long as it is tempered with an understanding of its sources and pitfalls. Moreover the observation that the mechanisms responsible for particular phenomena are local and varied does not by itself imply that we cannot say some quite useful things about what mechanisms, in general, are. This book is meant to show that there is indeed something useful that can be said generally about the character both of nature and of science. Thinking about mechanisms can help show how these things, in the most general sense, hang together.

2

Mechanisms

Most scientific pursuits can be explained as a search for "mechanism." Though the definition of mechanism almost universally corresponds to gaining insight into the what, how and why of the chosen research subject, the specific criteria for establishing a mechanistic hypothesis vary significantly among disciplines.

(*Nature Chemical Biology*, "The Mechanistic Meeting Point," 2007)

Use of the word "mechanism" is common across the whole range of scientific and technological discourse. Physicists, chemists, biologists, psychologists, sociologists, and historians speak of mechanisms, as do engineers, pharmacologists, and epidemiologists. A search of a recent month of the journal *Nature* gets over a thousand hits. Mechanism talk is clearly important to science, but it is reasonable to ask whether all of this mechanism talk is really talk about the same thing. When a biochemist talks of the mechanism of protein synthesis, an ecologist discusses the mechanism of competitive exclusion, and an economist speculates on the mechanisms responsible for a decrease in bank lending, is there anything non-trivial that these mechanisms have in common that entitles us to think that mechanisms form a conceptually interesting category? A central thesis of this book is that they do.

We must honor both the commonalities among and differences between the varieties of mechanisms. To do so I will start in this chapter and the next with a broad and encompassing characterization of what mechanisms are and how they are represented. This will be followed, in Chapters 4 and 5, with an account of mechanism kinds and a taxonomy of types of mechanisms that will attempt to tease out the differences between the various kinds of things that are called mechanisms, and the implications of those differences for the methods we use to investigate, describe, explain, and manipulate them.

I will begin with a terse characterization of what mechanisms in general are, which I call **minimal mechanism**:

A mechanism for a phenomenon consists of entities (or parts) whose activities and interactions are organized so as to be responsible for the phenomenon.

This characterization is minimal in the sense that little is required for something to be a mechanism. Consequently, minimal mechanisms will be widespread. There are two reasons I begin with a minimal characterization. The first concerns descriptive adequacy. The characterization needs to be broad enough to capture most of the wide range of things scientists have called mechanisms, and thereby to understand resemblances between mechanisms across the sciences. The second concerns ontological adequacy. If the argument in this book is correct, mechanisms in this minimal sense constitute the causal structure of the world. Mechanisms must be everywhere causes are, and that is pretty much everywhere.[1]

Readers familiar with the New Mechanism literature will be aware that the question of just how one should characterize or define the term "mechanism" has been much discussed, and that my minimal mechanism characterization is not de novo. It follows closely a recent proposal from Illari and Williamson, that "a mechanism for a phenomenon consists of entities and activities organized in such a way that they are responsible for the phenomenon" (2012, 120). In their paper, Illari and Williamson offer up their characterization as a response to three "main contenders" in the literature. It will be useful to have these earlier characterizations in view:

Mechanisms are entities and activities organized such that they are productive of regular changes from start or set-up to finish or termination conditions.

(Machamer, Darden, and Craver 2000, 3)

A mechanism for a behavior is a complex system that produces that behavior by the interaction of a number of parts, where the interactions between parts can be characterized by direct, invariant, change-relating generalizations. (Glennan 2002a, S344)

A mechanism is a structure performing a function in virtue of its component parts, component operations, and their organization. The orchestrated functioning of the mechanism is responsible for one or more phenomena. (Bechtel and Abrahamsen 2005)

Illari and Williamson's proposal and my minimal mechanism draw on elements of each of these characterizations, and, terminological differences aside, capture what they and I believe represents a consensus among philosophers identified as New Mechanists concerning necessary conditions on what should count as a mechanism.

In claiming a consensus, I do not want to overstate the agreement on these matters. For one thing, people differ on how exactly to cash out the various concepts appealed to in minimal mechanism. More importantly, it is sometimes argued that the different projects of interest to New Mechanists require different conceptions of mechanisms. Andersen (2014a; 2014b), for instance, has claimed that the sort of permissive characterization of mechanisms I have given here is best suited to addressing a set of metaphysical issues about the ontology and causal structure of the world, while a proper philosophical understanding of mechanistic science requires a more restrictive conception—one that insists that mechanisms are to some degree regular.[2]

---

[1] I hedge here by saying "pretty much everywhere" because there are complicated questions about whether or in what sense causes and mechanisms can be found in the domain of fundamental physics. I discuss these matters in Chapter 7. See also Kuhlmann and Glennan (2014); Kuhlmann (2018).

[2] Arnon Levy (2013) has, in a related argument, distinguished three kinds of New Mechanism—Causal, Explanatory, and Strategic. Like Andersen, he distinguishes a weaker notion (Causal Mechanism), which

While there is clearly value in distinguishing these philosophical projects, this does not make them unconnected. If Andersen's analysis is right, the more restrictive scientific sense of mechanism (which she calls mechanism$_1$) simply adds an additional condition to the set of necessary conditions that I have characterized as minimal mechanism (which she calls mechanism$_2$). Moreover, to the extent to which successful mechanistic science must refer in some way to mechanisms as things in the world, methodological and metaphysical issues will inevitably be entwined. As Machamer, Darden, and Craver argued in their 2000 paper, a proper characterization of mechanisms must meet standards of descriptive, epistemic, *and* metaphysical adequacy. In the end, though, minimal mechanism is an ontological characterization of what mechanisms are as things in the world, rather than a semantic characterization of what the term "mechanism" means, either in scientific, philosophical, or commonsense discourse. While a great many scientific and commonsense uses of the term "mechanism" will refer to things in the world that count as mechanisms in the minimal sense, this will not and need not always be the case.[3]

My strategy in this chapter will be to flesh out the various characteristics of minimal mechanisms—to say, for instance, what count as phenomena, entities, activities, and interactions, or mechanistic organization. This chapter will emphasize what mechanisms are as things in the world and will try to say what features all (minimal) mechanisms share. In subsequent chapters, I will move in two different directions. First, I will shift from nature to science, in the sense that I will turn from questions about what mechanisms are to questions about how we discover and represent them, and how mechanistic models figure in our scientific practices. Second, I will move from questions about what all mechanisms share in common to questions about the features that lead us to identify mechanisms as being of different kinds.

2.1 Entities and Activities

Mechanisms are compounds, in the sense that they are not simple but are composed of simpler things. New Mechanists agree that mechanisms are comprised of constituents of two sorts, which have variously been called parts and interactions (Glennan 1996), entities and activities (Machamer, Darden, and Craver 2000), and components and operations (Bechtel and Abrahamsen 2005). These variations are largely terminological,

serves a more metaphysical purpose, from more restrictive notions (Explanatory and Strategic Mechanism), which are more closely connected to the methods of mechanistic science.

[3] Andersen (2014a), for instance, identifies five senses of mechanism, only two of which she associates with New Mechanism. While the term "mechanism" is used in all of these senses, it is clear that at least a couple of them are about things that are not minimal mechanisms. Similarly, it is possible to find references to mechanisms in the scientific literature that do not meet the criterion of minimal mechanism. For instance, Goodwin (2018) observes that the term "reaction mechanism" is typically understood by organic chemists to refer to a certain kind of model—not to a mechanism, qua worldly entity. Or, in quantum physics, it is not clear that the Higgs mechanism (which is thought to explain the mass of certain subatomic particles) would count as a minimal mechanism, because, among other things, it is not clear whether it includes anything like an "activity."

and I will usually use the terms more or less interchangeably as context dictates.[4] An important part of the mechanistic consensus is the idea that mechanisms must have *both* these constituents. Mechanisms are made of parts (entities), but these parts must *act* individually and collectively in order to produce the phenomenon for which a mechanism is responsible.

Grammatically, entities are referred to by count nouns (or sortals). Examples of entities include proteins, organisms, congressional committees, and planets. Entities are, in a sense we will explore later in this chapter, objects—things that have reasonably stable properties and boundaries. Activities on the other hand are referred to by verbs. They including anything from walking to pushing to bonding (chemically or romantically) to infecting. They are a kind of process—essentially involving change through time. In most or perhaps all cases, the entities and activities that constitute mechanisms are themselves compounds. Entities like proteins or organisms are composed of further entities, their parts, and activities like bonding or infecting are processes that are composed of further activities.[5]

Entities and activities are not abstract; they are fully determinate particulars located somewhere in space and time; they are part of the causal structure of the world. Sometimes there are abstract structures that can be characterized with mechanistic metaphors—but they are not mechanisms. Take, for instance, a Turing machine. A Turing machine provides a constructive procedure for computing a function, but the machine qua abstract instruction set is not a physical thing at all. It could of course serve as an idealized description or model of a concrete physical device (idealized because, among other things, the physical device would not have an infinite tape). If one finds or builds such a device, that device would be a mechanism, and the Turing machine would be a model of it. But the mathematical structure is not a thing in the world; it has no causal powers; it isn't a mechanism (cf. Piccinini 2007a).

There cannot be activities without entities or entities without activities. This is why, in my characterization of minimal mechanism, I use the possessive pronoun in my characterization of minimal mechanisms: "entities (or parts) *whose* activities and interactions..." The first direction is obvious. Activities must have actors; if there is pushing, there must be something that is doing the pushing and something that is

---

[4] The terminological choices are vexed. Each of the proposed terms comes with baggage that can obscure the issues at hand. This seems particularly challenging with respect to the words referring to entities/components/parts. The term "part" is too suggestive of part/whole relationships of mereology, which do not fully capture the relationships between mechanisms and their constituents. The term "component" has too many connotations of engineering and design. The term "entity" is the vaguest, and in this sense may be the best, but it has one decided disadvantage: within metaphysics the term "entity" is used to refer generically to any member of the ontological zoo—so among the entities we might believe in are events, substances, properties, processes, tropes, and so on. It is thus a far broader category than the New Mechanist's entities. But since the use of "entity" has now become deeply entrenched in the mechanisms literature, I will accede to that usage. When I want to refer more generically to ontological categories, I will use the words "constituents" and "things."

[5] There may be some set of fundamental entities and activities that do not themselves have parts. See Sections 2.4 and 7.5 for discussion.

being pushed. But equally, there are no entities that are not active, at least potentially. To see this, consider an example of an entity that does not look very active, say, a rock sitting upon the ground. In the first place, the rock can potentially be used in all manner of mechanisms. It could, for instance, be picked up, placed in the sling of a slingshot, and launched so as to break a window. But even the rock, just sitting there, has an active part in certain mechanisms. For instance, the rock reflects light; it redirects wind and sound waves; it exchanges thermal energy with whatever it is in contact with.[6]

In my characterization of minimal mechanism, I have used the phrase "activities and interactions" rather than simply "activities." Why do we need both? The reason has to do with the arity of activity relations. As I understand it, the term "interaction" implies some relation between more than one entity (actors, agents, patients), while some activities can be done on their own. Daydreaming is an activity, but not an interaction, while having a conversation is an interactive activity. As Illari and Williamson (2012) point out, the principal descriptive advantage of the term "activity" is that it transparently allows for unary activities to be included within the description of a mechanism. Consider for instance a mechanism by which a courier carries a message from a sender to a receiver; imagine a messenger on the battlefield who is dispatched by the colonel to tell the general what is happening on the front. The courier talks with the colonel (a binary activity); the carrier then runs to find the general (a unary activity), and upon finding her, tells her what the colonel wished to report (another binary activity). A description of the mechanism that simply indicates that the colonel interacts with the courier, who interacts with the general, misses the fact that there is temporally extended non-relational activity that accounts for how the message is transmitted. This shows that we cannot simply speak of entities (or parts) and interactions. But why then can we not dispense with the term "interaction"? The reason, as Tabery (2004) has argued, is that proper mechanisms require that at least some of the parts interact. For some entity that is part of a mechanism to actually contribute to the production of the phenomenon for which the mechanism is responsible,

---

[6] MDC's use of a conjunction "and" in the phrase "entities *and* activities" appears to be motivated by their desire to avoid what they called "entity-bias" (or "substantivalism")—a problem that they attribute to earlier analyses of mechanisms, notably Bechtel and Richardson (1993) and Glennan (1996). They instead argue for what they call "dualism"—the doctrine that any mechanism requires both entities *and* activities, and that neither category is reducible to the other. I believe the claim that our analyses were guilty of entity-bias does not stand up to scrutiny (Glennan 2008; cf. Tabery 2004) and that there is not now nor ever was any disagreement with the claim that there cannot be entities without activities or activities without entities. Nonetheless, I think the term "dualism" is unfortunate because it implies (in ways reminiscent of Descartes) that activities and entities are ontologically distinct. MDC rightly point out that it is frequently useful to characterize activities independently of their entities or entities independently of their activities, but this fact does not imply (and MDC do not suggest it implies) that entities can exist without activities or activities without entities. The possessive pronoun in the phrase "entities whose activities" is more clearly suggestive of this interdependence, though it too has its problems; the grammatical relation of possession could be taken to suggest the ontological primacy of entities. Perhaps there is an entity bias in the grammatical forms of English—one can naturally speak of the dog's walking while one cannot naturally speak of the walking's dog—but these linguistic challenges do not detract from the fundamental point of ontological agreement among the New Mechanists.

it must produce changes in other parts of the mechanism; there is no production without interaction. That said, I shall understand the category of activity as MDC do, to be inclusive of interactions. Properly but pedantically, we could have phrased our requirement "entities whose activities (at least some of which are interactions)..." For many purposes in this book it will be useful to speak of the broader category of activities, while in other cases we will need to focus on interactions between parts.

2.2 Organization, Phenomena, and Responsibility

According to minimal mechanism, a mechanism's parts must be so organized that their activities and interactions are, collectively, responsible for some phenomenon. But what is it for entities and activities to be organized, what counts as a phenomenon, and what is this relation of responsibility? These questions are related, and a useful starting point for discussing them is to consider a kind of representation of a mechanism that is due to Carl Craver, which I will call a Craver diagram (Figure 2.1).

Craver diagrams represent mechanisms as organized in two dimensions. There is, first, a "horizontal" dimension, which is spatio-temporal and causal. The parts of the mechanism (represented in the center of this diagram by the X_is) engage in activities and interactions (represented by φ_is and connecting arrows). The second, "vertical" dimension, generally called mechanistic constitution (Couch 2011; Leuridan 2012; Povich and Craver 2018), is the relationship that holds between a mechanism as a whole and the collective organized activities and interactions of its parts. The mechanism as a whole is a complex entity or system S engaged in some activity ψ. The mechanism's activity, S ψ-ing, is said to be constituted by the organized activities and interactions of its parts (X_is φ_i-ing). As the Craver diagram indicates, the constitution relation is

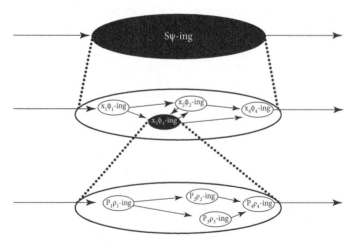

Figure 2.1 A Craver diagram (reproduced from Povich and Craver 2018).

multi-level. The X_is that are φ_i-ing are at once acting components of the mechanisms they constitute, and are constituted of further acting components, the P_is that are ρ_i-ing.

Take animal digestion as an example. At the top of the mechanistic hierarchy, an animal is a system S, and digesting is an activity ψ that the animal engages in. The animal has parts—for instance, the mouth, esophagus, stomach, pancreas, and small and large intestines. These parts (the X_is) engage in various activities and interactions (the φ_i-ings) like chewing, swallowing, secreting of enzymes. The activities of these parts are themselves mechanistically constituted by the activities of their parts, and the mechanism as a whole, the digesting animal, is in turn situated within larger ecological mechanisms. In this picture, the mechanism's phenomenon is some activity ψ of the system S, here the animal digesting. The phenomenon is what the mechanism does. The mechanism itself is what is responsible for that phenomenon; it is the "how it does it." This responsibility typically involves both the horizontal and vertical dimension—a mechanism's parts and their activities and interactions both cause and constitute the mechanism's phenomenon.

Mechanisms behave as they do because of the organized activities and interactions of their parts. The purport of the term "organized" is simply to indicate that a mechanism's phenomenon depends not just on what the parts of the mechanism are, or on what activities those parts engage in, but on how the parts and their activities are arranged. Mechanisms are not just heaps of parts. To take a very simple example, the parts of an engine will not work as an engine unless the engine's parts are carefully arranged and connected. Similarly, the behavior of the animal's digestive mechanism depends upon how each of the parts is physically connected to each other, on the timing with which food passes through the system, on neural and endocrinological signals that impact the secretion of enzymes, etc. In a Craver diagram, organization is expressed primarily via the arrows between entities. These arrows remind us that the primary form of organization in a mechanism is causal. Other forms of organization, from basic spatial and temporal organization in physical, chemical, and biological mechanisms to social and cultural organization in the mechanisms described by the social sciences, are important just because they promote, inhibit, and guide causal interactions.

According to minimal mechanism, all mechanisms are mechanisms for some phenomenon. This principle is consistent with common usage in the sciences, where it is commonplace to identify mechanisms by what they do or what they produce— digestive mechanisms, mechanisms of protein synthesis or gene expression, speciation mechanisms, mechanisms for long-term memory, or market-clearing mechanisms. A phenomenon is what is used to identify and delimit its mechanism (Bechtel and Abrahamsen 2005; Craver 2006; Darden 2008; Glennan 1996). If we characterize a mechanism as a thing S (entity or system) that is ψ-ing, what counts as part of the mechanism's phenomenon (the ψ-ing) depends upon what is actually contributing to the production of the phenomenon. Take for instance a car. A car is a system with many mechanisms, but pre-eminently a car is a system for driving on roads. While the car has many parts, not all parts contribute to its driving. The wheels matter but the rear-view mirror does not. Those parts of the car that are implicated in the mechanism are its *working parts* (Bechtel and Abrahamsen 2005; Darden 2005; Craver 2007). A similar point may be made about the activities involved. Parts may do more than one thing, and only some of those things will contribute to the production of the phenomenon. The engine, for instance, is a working part of the car's mechanism for turning the wheels, but it produces heat as it rotates the crankshaft. The rotation, but not the heat, is a *working activity* in the mechanism.

There is a close relationship between the concept of a mechanism's phenomenon and the idea of function, but not all of the things that meet the criteria for minimal mechanisms will have functions (see Section 5.1.3). If a mechanism is designed for some purpose, or if it has acquired its characteristics as a result of a process of natural selection, then the mechanism's phenomenon can indeed be a function; digestion is indeed the function of the digestive system, and the principal function of the car is to move about on roads. But in other cases we have mechanisms that are responsible for phenomena, which are not functions in any teleological sense. On the one hand, mechanisms designed (or selected) for other things can have side effects that are not part of their functions. The engine produces heat, and there is a mechanism responsible for it, but producing heat is not the function of the mechanism. Also, there can be phenomena that are produced by mechanisms that have come to be independently of any design or selection process. Volcanoes, for instance are mechanisms for spewing lava, but there is no hint of design or function in their eruptions.

What I have called the mechanism's phenomenon could also be called its behavior, and we also speak of the patterns of phenomena/behavior for which a mechanism is responsible. These expressions share an important semantic feature; they have something to do with the visible or the apparent. Phenomena, behaviors, and patterns are, literally or metaphorically, seen or observed. When we say that a mechanism is responsible for a phenomenon (or pattern or behavior) we are saying that there is something behind that phenomenon. This distinction between the phenomenon and the mechanism that is responsible for it, or the pattern and the process that produces it, is central to understanding the nature of mechanisms and their importance for the scientific enterprise. The ecologist Simon Levin makes just this point:

Theoretical ecology, and theoretical science more generally, relates processes that occur on different scales of space, time, and organizational complexity. Understanding patterns in terms of the processes that produce them is the essence of science, and is the key to the development of principles for management. Without an understanding of mechanisms, one must evaluate each new stress on each new system de novo, without any scientific basis for extrapolation; with such understanding, one has the foundation for understanding and management.   (Levin 1992, 1944).

The sorts of phenomena that mechanisms are responsible for include not just the kind of regular or recurrent phenomena we have discussed so far, but also non-recurrent or "one-off" phenomena. The continual beating of a heart is a phenomenon just as is the one-off occurrence of a heart attack. In addition, while we have focused so far on systems acting—like cars moving or animals digesting—we can also count among phenomena capacities to act, even if those capacities are not manifested. For example, many octopus and squid have remarkable abilities to change color in order to camouflage themselves as they move through their environment. We can characterize the mechanism that is responsible for this capacity, apart from any particular manifestation of it.

The examples in this section will, I hope, give some sense of the rich variety of phenomena for which mechanisms can be responsible, as well as the basic ways in which mechanisms can be organized so as to give rise to these phenomena. These concepts will be fleshed out further in the remainder of this chapter, but a fuller treatment will have to wait for Chapter 5, where I will offer a more detailed and systematic account of the different kinds of phenomena and organization.

2.3 Mechanisms, Systems, and Processes

In the previous sections, I have on occasion helped myself to the terms "system" and "process" as I have developed my preliminary characterization of mechanisms. In this section, I would like to get clearer on what systems and processes are, and how they are related to mechanisms. Like the term "mechanism," the terms "system" and "process" are used in many ways by scientists and ordinary folk, and these terms have also been the subject of philosophical scrutiny.[7] Without pretending to offer any definite analysis of these concepts, we can acknowledge some platitudes about common usage of these terms. Systems seem to have two important features: first, systems are wholes that are made up of parts; second, systems do things, with a system as a whole doing things in virtue of the activities and interactions of its parts. A good example of a system is the United States Postal Service. It has many parts—post offices, mailboxes, mail trucks, clerks, etc.—and the activities of these parts allow the system to do things: delivering mail, of course, but also selling stamps, accepting passport applications, and so on. Another example is an ecosystem, which consists of many kinds of parts—both biotic parts like plants and animals, and abiotic ones like soils, nutrients, water, and air. These parts interact in various ways, cycling nutrients and energy around the system. While the things that systems do take time, systems seem to be characterized chiefly by their part-whole organization and their persistent causal structure.

Processes too are compounds, but their primary mode of organization is temporal and causal. A process, like the evolution of a star or the development of an infant into

---

[7] In philosophy, discussions of processes have typically centered around metaphysical questions about the relative merits of substance and process ontologies (Whitehead 1929; Rescher 1996). There has not to my knowledge been much extended philosophical discussion about the ontological features of systems, though the concept of a system has been an object of theoretical analysis and interdisciplinary research in fields such as cybernetics and general systems theory (von Bertalanffy 1950; Wiener 1950). Recently John Dupré (2012) has argued for processual ontology for the life sciences.

an adult, is characterized by its stages.[8] In contrast to systems, it is not essential to our concept of processes that they be regular or repeatable. There are many regular and repeatable processes, like processes we observe in cellular metabolism or in plant and animal development, but it is also possible to find and characterize processes that are one-off. These sorts of processes are just causal chains involving sequences of activities and interactions between entities leading up to singular and unrepeatable events—from chance meetings in cafes, to asteroid collisions or genetic mutations.

Going forward, I will speak of mechanistic systems and mechanistic processes. A mechanistic process just is a compound process of the kind described in the previous paragraph. To call a process mechanistic is to emphasize how the outcome of that process depends upon the timing and organization of the activities and interactions of the entities that make up the process. A mechanistic system, on the other hand, is a system that regularly engages in or is disposed to engage in mechanistic processes. So, for instance, a mail system is a mechanistic system, because it regularly engages in mechanistic processes like delivering the mail. Notice that a mechanistic system is not, strictly speaking, a mechanism; it is rather a thing in which mechanisms act. Using Craver's terminology, S is a mechanistic system (an animal, a car, etc.), and when that system S does something ψ (digesting, driving, etc.), the mechanism is the organized activities and interactions of entities within the system that is responsible for that ψ-ing.

One reason to distinguish a system from the mechanisms that act within it is that systems are not always acting, and mechanistic phenomena always involve the occurrence of activities and interactions (Kaiser and Krickel 2016). Systems are in fact a species of the category we call entities (or objects)—entities that are compound rather than simple. They have parts that interact with each other in organized ways that give the entity the capacities that the entity has. These we can call *mechanism-dependent capacities* of systems.[9] Systems, as entities, can also be parts of further mechanisms. In Figure 2.1, the X_is are systems just as much as S is.

Another reason not to equate systems with mechanisms is that systems typically act in many ways and depend upon many mechanisms. A mouse, for instance, is a system that does many things: it eats food and produces waste; it regulates its body temperature; it moves about mazes; it copulates with mouse-mates. If we consider any physiological

---

[8] It is worth remarking that the notion of process I am characterizing is distinct from another notion familiar in the causation literature. The notion of a causal process in process theories of causation (especially those of Salmon and Dowe) is of a world line of an object. An asteroid moving through space is a causal process in this sense. My sense of causal process assumes that processes are compounds involving different parts interacting over time. The relationship between process theories of causality and mechanisms will be discussed in Chapter 7.

[9] This is obviously a permissive conception of a system. It makes, for instance, a broken bit of branch, with its capacity to absorb moisture, a system. The honorific "system" I expect most often applied to compound entities which were designed for or selected to perform certain functions, and which have a certain degree of organizational complexity—like postal systems or digestive systems—but as with mechanisms, I see no clean line to draw between the more and less systematic compound entities.

or behavioral activity or capacity of a mouse, the mouse has a mechanism that is responsible for it, but the plain mouse is not a mechanism.[10]

Even so, sometimes it is natural to speak of a system as a mechanism for one particular thing. This is most evident in cases of designed objects—refrigerators are mechanisms for keeping things cold, and can openers are mechanisms for opening cans—but the same can be said for systems that have selected functions, like hearts and stomachs. There is no harm in this talk so long as it is remembered that any system will have many behaviors and capacities that will depend upon different mechanisms.

There is an asymmetry between mechanistic systems and mechanistic processes: mechanistic systems, when they act, engage in mechanistic processes, but not all mechanistic processes are actions of systems (Glennan 2002a). For instance, kidneys are systems that (among other things) remove wastes from blood and concentrate them in urine (a process). What makes these mechanisms systematic is their stable organization and regular and repeated operations. The postal system would not be a system unless it was organized so as to repeatedly engage in processes that produce its behaviors—processes like the delivery of letters and packages or the processing of passport applications. On the other hand, there are certainly processes involving entities that engage in activities and interactions that produce some phenomenon, but where that process is not systematic or repeatable. To the extent that these mechanical processes are not systematically organized, they are instances of what I call ephemeral mechanisms (Glennan 2010b). Historical explanations of one-off events (extinctions, marriages, collisions between passenger liners and icebergs, and so on) involve the description of such mechanisms.

In contrast to one-off ephemeral mechanisms, the processes systems engage in are regular, in the sense that their behavior recurs and the mechanical system operates "always or for the most part in the same way" (Machamer, Darden, and Craver 2000, 3). Relatedly, mechanistic systems are robust, in the sense that their behavior is relatively insensitive to changes in background conditions. Robustness and regularity, though, are matters of degree. Robustness of processes is connected to systemhood because whether or not a set of entities are engaged in activities and interactions that form a robust process is an important criterion for judging whether the set of entities so engaged forms a system.[11]

---

[10] Formally speaking it would be possible to conjoin many activities into one complicated activity—so that S might be a mechanism for the compound activity "$\psi_1$ing and $\psi_2$ing and ..." I doubt that there are completely objective ways to count the number of activities that a system engages in; pragmatic and explanatory factors will enter in. However, it is clear at least as a descriptive matter that we treat many complex entities (i.e., systems) as having multiple capacities and activities in which they engage.

[11] The degrees and respects in which mechanisms behave regularly has been a subject of debate, which I discuss in Section 2.6, and again in Chapter 5. I borrow the term "robust process" from Sterelny (1996). See Wimsatt (2007, ch. 4) for a classic discussion of robustness and robustness analysis. As Calcott (2011) explains, Wimsatt has several different notions of robustness in play in his work—some of which are epistemological and methodological while others are ontological. Robustness of phenomena and processes is an ontological notion, though, as Sterelny emphasizes, robustness is in part a function of descriptive grain.

The system/process distinction is crucial to understanding the different kinds of relationships between mechanisms and their phenomena. Darden and Craver have recently distinguished between mechanisms that produce phenomena and mechanisms that underlie phenomena. Writing about the process of mechanism discovery, they characterize the distinction this way:

In the case of a productive mechanism, one typically starts with some understanding of the end product and seeks the components that are assembled and the processes by which they are assembled and the activities that transform them on the way to the final state. In the case of an underlying mechanism, one typically breaks a system as a whole into component parts that one takes to be working components in a mechanism, and one shows how they are organized together, spatially, temporally, and actively such that they give rise to the phenomenon as a whole.   (Craver and Darden 2013, 65–6)[12]

Craver and Darden schematically represent these relationships using two different kinds of diagrams, redrawn in Figure 2.2.

The underlying relationship is represented using a simplified Craver diagram, with the vertical levels representing mechanistic constitution. Production, on the other hand, requires no ascent to a higher level. It represents only the horizontal dimension of spatio-temporal and causal organization. As I understand it, then, mechanistic processes that underlie phenomena are always mechanisms operating within systems.[13]

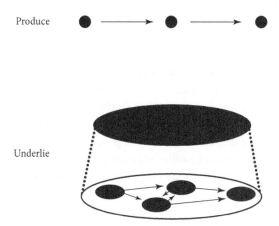

Figure 2.2   Varieties of mechanism-phenomenon relations (reproduced from Craver and Darden 2013, 66).

---

[12] Craver and Darden actually identify three kinds of mechanism-phenomenon relationships—mechanisms may either produce, underlie, or maintain phenomena. A maintenance mechanism maintains the state of some system in the face of perturbations for its environment. While this is a very important class of mechanisms biologically (think of mechanisms that maintain balances of metabolites or body temperature), I see mechanisms that maintain phenomena as a species of system mechanisms.

[13] I shall have more to say in Chapters 6 and 7 about the notion of causal production. There we will see that there can be process mechanisms in which there can be causal dependencies that are not strictly speaking productive.

The relationship between activities and processes is analogous to the relationship between entities and systems. Indeed, most activities just are mechanistic processes, according to the sense of the term "process" I have sketched above. Processes take time, and bring about changes in the entities involved in them. There is a process by which proteins unfold, a process by which eagles select mates, a process by which a speech sound is produced or recognized. So these activities are processes, and these processes depend upon mechanisms.

Let me close this section by noting some important questions that remain insufficiently answered. The first concerns the nature of the relationship I have called mechanistic constitution. I hope it is clear enough what relation I am pointing to— the relationship that obtains between whole mechanisms and their constituent entities and activities—but I do not think that in pointing to this relation or giving examples of it, we have yet given an answer concerning the nature of this relationship. Metaphysically, there are a number of different options that have been offered to characterize the vertical relation between mechanisms and their constituents, including realization, grounding, and identity. It is not clear, which if any of these will do, or if some different account is needed. Another issue concerns the relation between mechanisms, activities, and causes. Beyond these metaphysical questions, there is a more practical question. If entities, activities, and the mechanisms they constitute are compounds, of what are they compounded? Where does one entity or activity or mechanism end, and when does another begin? And on what account do we decide that a collection of interacting entities is to count as a whole mechanism? Put in terms of Craver diagrams, when do we put a circle around a collection of X's and call it an S? Indeed, one might wonder whether there is a fact of the matter about such a question, or whether the whole question should be interpreted as a practical/instrumental one.

In the remainder of this chapter, I shall begin to answer these questions, first by offering further explanation of the character of entities and activities, and the relation I shall call mechanism-dependence, and second by offering some suggestions on how, practically, one draws boundaries between entities, activities, mechanisms, and their environment. This will allow me to offer a characterization of the ontology that falls out of New Mechanical Philosophy. Some questions will, though, remain; I will attempt to address these in the balance of the book.

2.4 Entities and Activities Elaborated

In Section 2.1, I provided an initial characterization of entities, activities, and interactions. Entities are things or objects that are parts of mechanisms, and activities and interactions are what those entities engage in. I am now in a position to fill these descriptions out and say something more about the nature both of entities and of activities, and to explain the ways in which entities and activities not only are constituents of mechanisms but themselves depend on mechanisms. I will begin with activities, and then turn to entities.

Probably the most important fact about activities is that there are a lot of them (Machamer 2004; Bogen 2008c). There is no one thing that is "acting," but are instead many kinds of acts. This position is closely connected to views about causality that have been championed by Elisabeth Anscombe. Anscombe writes:

The word "cause" can be *added* to a language in which are already represented many causal concepts. A small selection: *scrape, push, wet, carry, eat, burn, knock over, keep off, squash, make* (e.g., noises, paper boats), *hurt*. But if we care to imagine languages in which no special causal concepts are represented, then no description of the use of a word in such languages will be able to present itself as meaning *cause*. Nor will it contain words for natural kinds of stuff, nor yet words equivalent to "body," "wind," or "fire." For learning to use special causal verbs is part and parcel of learning to apply the concepts answering to these.   (Anscombe 1993, 93)

Anscombe's specific causal concepts ("scrape," "push," etc.) refer to the sorts of things that New Mechanists have called activities. Anscombe's point is in the first instance a conceptual and epistemological one—that particular causal concepts are psychologically, epistemologically, and semantically prior to general concepts like "cause," "interact," and "produce." But Machamer and Bogen draw an ontological consequence: There is no one thing in the world as "causing," "producing," or "acting," but only a variety of specific sorts of activities.

Peter Godfrey-Smith sees the sort of view championed by Anscombe, Machamer, and Bogen as a species of causal pluralism that he calls causal minimalism. He offers the following succinct formulation:

1. "C was a cause of E" is true iff the relation between C and E can also be described using some member of set S, or can be described as a chain of relations each of which can be described using some member of S.
2. S is a set of causal verbs and other linguistic formulas which represent "special causal concepts" in Anscombe's sense. (Godfrey-Smith 2009b, 333)

If we rephrase Godfrey-Smith's formulation slightly in activities language, we can say that "C interacts with E" is true if and only if the interaction between C and E can also be described using causal verbs characterizing specific interactions or chains of interactions relating C to E (where C and E are now entities rather than events). If it is true that I interacted with a bee, for instance, it must be true because of some specific sort of interaction. The bee may have stung me, for instance, or I may have just seen it and moved away.

Is there something in general that we can say about what unifies the set of things we call activities? Machamer doubts it:

Activities have criteria that can be given for their identification. We know how to pick out activities such as running, bonding, flowing, etc.…One might try to do something more general by giving the conditions for all productive changes, but then one would have to find out what all producings have in common and by what are they differentiated from non-producings. It is not clear that they all have any one thing in common or are similar in any significant way, but neither commonality nor similarity are necessary conditions for an adequate category.   (Machamer 2004, 29)

Put in Godfrey-Smith's terms, Machamer is suggesting that there is no unifying concept that determines the extension of the set S of activity words.

I think Machamer's pessimism is unwarranted. We can grant the thinness or abstractness of concepts like "cause" and "activity" without thereby denying that there are some objective features that distinguish causes and activities from non-causes and mere happenings. Part of what we must do to see this is to separate conceptual from ontological issues. It may be that there is at best a set of family resemblances between our activity concepts—but this does not mean that there are not properties that activities (as things in the world) all share. This is important, because if activities are what ground causal relations, there must be something about activities that makes them what they are.

The following are some important shared characteristics of activities (including interactions), which we can take as necessary conditions:

A1.   Activities require entities (parts, components) to act and be acted upon.

A2.   Activities produce change in entities (parts, components) that act or are acted upon.

A3.   Activities manifest the powers (capacities) of the entities involved in the activity.

A4.   Activities are temporally extended processes.

A5.   Most or all activities are mechanism-dependent.

A full understanding of these conditions will require us to delve into the relationship between activities, interactions, and causality, which will occupy us in Chapters 5 and 6, but we can clarify several points immediately.

First, regarding A1: With respect to activities that are interactions, we often distinguish between the active "doer" of the activity and the passive object of the activity, but this distinction is not a deep one, in the sense that all actors in an interaction produce changes in other actors, and often the active/passive distinction is a matter of degree (cf. Heil 2012, sec. 6.3). As an example, consider the activity of eating. If Franz eats the Strudel, this is an interaction between Franz, the actor, and the Strudel, the object of the interaction. And while it is true that Franz is doing something to the Strudel, the Strudel is also doing something to him. To take another example, as a waterfall strikes the rocks at the bottom, the waterfall over time will erode the rocks, but so too the rocks will change both the motion of the water and the materials flowing within it.

Regarding A2, notice that *both* monadic activities and polyadic interactions are change producers. It is the sine qua non of interactions that they produce changes in at least some of the interactors. When two billiard balls strike each other, the strike produces changes in the momentum of each; when Sisi kisses Franz, the kissing produces changes in both Franz and Sisi; when the Federal Reserve Board sets the interest rate, it changes the behavior of banks. Monadic activities still produce change, but it is not one thing changing another, but something changing itself. Take for instance my daydreaming. This daydreaming is something that I alone do, but when I daydream, this involves some psychological and neurological processes occurring in parts of my body and central nervous system, and this process will produce changes in those parts and thereby produce changes in me.

Criterion A3 signals the close relationship between activities and powers or capacities. The central difference between activities and powers is that activities are actual doings, while powers express capacities or dispositions not yet manifested. For instance, a battery may have the power to generate current without actually generating current (which is an activity). This criterion also reminds us that activities are powerful—that they are things that an entity does, not merely something that happens to an entity.

Criterion A4 reminds us that all activities are temporally extended processes. This stipulation applies both to monadic activities and to interactions. While I do not know of anyone who would deny the temporally extended character of interactions, this feature is often lost in philosophical treatments. For instance, Salmon's approach to causation relies upon a deep distinction between processes and interactions. Processes are temporally extended transmitters of causal influence (e.g., a ball in motion heading towards a window), while interactions are points of intersection between these processes (e.g., where the ball process intersects the window process). It will be crucial to our account of interactions (like the interaction between the ball and the window) that these points of intersection are not extensionless points, but take time.

Criterion A5 highlights the most important feature of activities, that they are generally mechanism-dependent. Bogen, who like Machamer is pessimistic about saying anything in general about activities, remarks that "What makes the idea that causally productive activities distinguish causal from non-causal sequences of events controversial is that there is no informative general condition which discriminates causally productive activities from goings on which are not causally productive of the effect of interest" (Bogen 2008c, 113). But, criterion A5 says something quite informative and general, namely that the productive character of activities comes from the productive relations between intermediates in the process, and that the causal powers of interactors derive from the productive relations between the parts of those interactors.

Bogen tells us that "according to Anscombe, it is a brute fact, for example, that the activity of pulling on a door can open it rather than coincidentally accompanying a door opening, that scraping a carrot removes rather than coincidentally accompanies the removal of its skin, that wetting something does not merely accompany its getting wet, and so on" (Bogen 2008c, 114). But is this fact really brute? In one sense, obviously not. It is hardly a brute fact that scraping a carrot removes its skin. The power of the knife to scrape the skin off the carrot depends upon the shape and hardness of the knife and the carrot, the pressure applied in the scraping, and so on. There is a mechanism by which the scraping is productive of the skin removal.

What Anscombe and Bogen are presumably getting at is that there is no general reductive criterion for distinguishing producings from non-producings. The account of mechanism-dependent production explains some producings in terms of other producings, not in terms of some non-causal features such as regularity, or counterfactual dependence. Also, Bogen is reminding us that our knowledge of some causal relations may be epistemologically basic. We can, *pace* Hume, observe door-pullings producing door-openings.

While the mechanism-dependence of (all or most) production is a unifying feature of productive activities, Bogen and Machamer are right to emphasize that there are no abstract and general criteria for identifying production. This follows from the non-reductive analysis of production whereby producings can only be characterized in terms of other producings. There is accordingly no such thing as THE ontology or THE epistemology of THE causal relation, but only more localized accounts connected with the particular kinds of producing.

But even in the absence of a reductive account, these general criteria for activities goes some way toward addressing what Godfrey-Smith and others have seen as a crucial worry for the activities approach (and causal minimalism generally). Godfrey-Smith's worry is that if there is nothing at all that characterizes causally productive activities as such, then we cannot meaningfully speculate about causal connections produced by as yet not understood activities. The criteria suggest some general conditions which these as yet undiscovered activities will have to meet to count as activities, and A5 in particular emphasizes that we will come to recognize heretofore unrecognized activities as genuine as we come to understand the mechanisms that underlie them, as well as the mechanisms in which they take part.

Finally, regarding A5, let me say something about the hedge that requires me to qualify that *most or all* activities are mechanism-dependent, rather than just saying that *all* activities are mechanism-dependent. The examples I have used in this chapter and indeed essentially all examples of activities discussed in the New Mechanism literature are clearly mechanism-dependent. There is a mechanism responsible for a protein's folding, or an animal's digesting a bit of food, or the match lighting when it strikes a rough surface. But one might reasonably wonder if there are a set activities or interactions that are essentially dependent on brute dispositions—ones for which no mechanism is responsible. Perhaps the nuclear forces acting on parts of atoms are an example of such. If so, these would be a set of fundamental activities or interactions on which other mechanisms depend, but which themselves are not mechanism-dependent. Alternatively, it is at least conceivable that there are mechanisms "all the way down." I will defer discussion of these alternatives until Chapter 7, but ultimately I will remain agnostic—hence the "most or all."[14]

Let me turn now to entities. While much has been made of the Anscombian argument for minimalism about causes, it has gone largely unnoticed that exactly parallel

---

[14] In the past (Glennan 1996; 2011) I have been less circumspect, arguing for a fundamental level. In Section 7.5 I consider both the physical and metaphysical considerations that motivate my current agnosticism as well as the implications of the different positions for a mechanistic account of causation.

arguments can be brought to bear in favor of minimalism about entities. We might say, imitating Anscombe, that

the word "entity" (or "part" or "component" or "object") can be *added* to a language in which are already represented many entity-concepts. A small selection: *ball, rock, orchestra, kangaroo, corporation.* But if we care to imagine languages in which no special entity concepts are represented, then no description of the use of a word in such languages will be able to present itself as meaning "*entity*."

The story we must tell about entities is accordingly similar to the one I have given so far regarding activities. Because the concept of entity is so thin, much of what can and should be said about entities—for instance about how they are detected, individuated, and classified—cannot be said about entities as such, but only about particular kinds of entities. But again, as with activity, it is possible to identify some general characteristics of entities, which serve as necessary conditions of entity-hood, and we need to offer an account of these characteristics in order to flesh out our account of mechanisms. This set of conditions parallels the list we have given for entities in the previous section:

E1.   Entities are what engage in activities and interactions.

E2.   Entities have locations in space and are stable bearers of causal powers (or capacities) over time.

E3.   The causal powers or capacities of entities are what allow them to engage in activities and thereby produce change.

E4.   Most or all entities are systems composed of parts and most or all of the powers of entities will be mechanism-dependent.

Condition E1 is the converse of condition A1 about activities. Entities are the actors that activities require. Condition E2 reminds us that entities are concrete objects with spatial locations. This principle should not be taken too strongly. Entities can have locations in the sense of E2 even if they are widely spread out and overlap other entities. There is no particular limit on how far an entity can be sprawled across space-time. Take, for instance, the faculty of my college. They are an entity formed out of the individual faculty members—faculty members who are currently spread out across the world. What unifies the faculty as a faculty is that they are stable bearers of causal powers—so the faculty are a faculty because collectively they have the powers to engage in activities that produce change (E3). For instance, the faculty have the power to change the curriculum, and this power will manifest itself in the lives of our students.

Condition E4 specifies that entities are mechanism-dependent. (The hedge "most or all" is added because of the considerations about fundamental entities discussed above.) To say of an entity that it is mechanism-dependent is to say more specifically that a (non-fundamental) entity S is a system with parts X, and the entity's capacities and activities are constitutively dependent upon the activities and interactions of those parts. This contrasts with the mechanism-dependence of activities, where a (non-fundamental) activity depends upon a process mechanism. While activities are primarily characterized by their stages and temporal/causal dependencies, entities are primarily characterized by their constituents, their boundaries, and their stable causal dispositions.

Condition E2 suggests that entities must be stable bearers of causal powers; the reason is that these causal powers are what allow the entities to engage in activities and interactions that allow them to be parts of mechanisms. Stability (and with it, entity-hood) is not an absolute characteristic, but should be understood relative to the timescales of the mechanisms in which the entities are involved. Living organisms are seas of change. Are they stable objects or dynamic processes? The answer depends upon the mechanistic context. Take for instance two mechanisms involving an animal. If the phenomenon under consideration is development, then the animal is a process in continual change. If the phenomenon under consideration is something more limited (say a social interaction between two animals) than each animal will be an entity that is part of a mechanism-dependent interaction.

To say that something is an entity is not to imply that it is inert or unchanging. Entities have parts, and the powers and activities of the entities will depend upon those parts and their ongoing activities and interactions. Stability often depends upon activity. This is nowhere more evident than in the case of living organisms. Organisms are paradigmatic entities, with stable properties, and they are actors in many ecological and evolutionary processes. But at the same time, organisms are open systems constantly interacting with their environment, exchanging matter and energy. As the environment changes, organismic response must adjust. The processes by which they maintain their properties (e.g., temperatures, concentrations of metabolites) are continuous and active, requiring the expenditure of energy gleaned from the environment. Organisms can become damaged and engage in self-repair. In short, while organisms may be entities, they are also hives of activity. Moreover, while organisms are systems that actively work to maintain the stability of their properties over time, they are by no means immune to change. In fact change is essential to their identity. Organisms have life cycles, sometimes of astonishing diversity and complexity, and their identity as organisms will not depend so much on the continuity of the parts of which they are made but the continuity of the processes by which they develop.

Stability also does not imply independence from environment. An object may be stable within the environment it occupies but change radically or be destroyed if its environment changes. An iceberg will, when the temperature stays below freezing, maintain mass and shape over long periods of time, but move it to a warm environment and it will start to melt, changing shape and losing mass.

For this reason, to say that an entity is real and stable is not to suggest that it is prior to or independent of the system of which it is part. The degree to which a part can exist independently of the mechanism it is embedded within will vary by the kind of mechanism. To see this, compare a component of a clock like a gear with a component of an animal like a heart. Both of these parts satisfy criteria E1–E3 so long as they remain embedded within their respective mechanistic systems, but there is a crucial difference. The gear is a component that came into being prior to the clock. It can be taken out of the clock and will still remain a gear. The heart on the other hand does not pre-exist the animal, but grew in the process of the animal's development. The heart also, except under exceptional conditions, will cease to be a heart when it is removed from the animal—in the sense that it will die and cease to have the capacities that it does.

2.5 Boundaries

Mechanisms have boundaries—starting points and ending points, insides and outsides—and so too do the entities, activities, and interactions that make them up. But it can be both empirically and conceptually challenging to decide what goes inside or outside, when a collection of entities or mechanisms constitutes a larger entity or mechanism, or where one mechanism ends and another begins. Following Franklin-Hall (2016), I will call this the carving problem, and my aim is to describe some of the principles that guide carving. The principal kinds of boundaries we will discuss are three. First, there are the boundaries of entities, objects, and systems (recalling the close connections between these concepts, articulated in Section 2.3). Second, there are the boundaries of processes (which include the boundaries of activities and interactions). Third, there are the boundaries of mechanisms themselves.

System/object boundaries that are of concern to our account of mechanisms are further divided into two kinds—first the boundary between a system and its environment, and second, the division of the system into parts. However, given the hierarchical nature of mechanisms, these kinds of boundaries amount to the same thing. A component X of a system S will itself typically be a system, so when we draw the boundary between X and its environment, we are at the same time drawing a boundary around a part of S.

In processes, in addition to the boundaries between entities that are engaged in a process, there are temporal boundaries between stages of that process. When does a process, an activity, or an interaction begin and end? As with object boundaries, we can see two kinds of process boundaries—the temporal boundaries between which the mechanical process as a whole starts up and terminates, and the temporal boundaries of activities and interactions within the process. And as with object boundaries, we see that the hierarchical nature of mechanisms implies that this distinction tends to collapse. The activities and interactions that are stages within processes are themselves embedded within larger processes, so that in saying when a mechanical system starts or terminates, we are also drawing stage boundaries within larger mechanical processes.

As a practical matter, it is often not too hard to carve a system into parts. If I disassemble my lawnmower, there are natural part boundaries that give me spark plugs, gas caps, fuel filters, etc. I could try to cut the mower into parts with different boundaries.

I might, for instance, be able to use a laser to cut my lawnmower into 1 cm cubes. But these are not the lawnmower's joints. It would require much more effort to carve it in this way, and it would be impossible to reconstruct a working mechanism from the pieces, since the cubes are not working parts of the lawnmower. While part boundaries are perhaps most obvious for engineered machines, similar observations hold for lots of naturally occurring systems. Turkeys really do have gizzards, hearts, feathers, wings, and legs. And it really is much easier to carve a turkey at its joints! But the fact that we seem to be able to rationally carve up many systems does not obviate the need for an account of the norms of carving. For one thing, a philosophical account of mechanisms and mechanistic explanation should illuminate and justify the implicit norms at work in our judgments about the intuitively obvious cases. For another, there are many contexts in which what the parts of a mechanism are is far from obvious, and the development of these norms could provide resources for judging the adequacy of proposed carvings.

Scientific realists have the hope that mechanisms in nature have joints that are there before we look, just waiting to be found. This hope may seem to be in tension with a principle often espoused by New Mechanists, which we can call the phenomenon-dependence of decomposition. As noted at the outset, mechanisms are always mechanisms for some phenomenon; but depending upon the phenomenon one seeks to explain, one will get different decompositions of a system into parts.[15] Systems, as has already been noted, can do many things—a car may be driven along highways, melt chocolate on its seats, spew exhaust fumes, and do many things besides. In Section 2.2, I observed that not all parts of a system will be working parts of the mechanism responsible for any particular activity of the system. The seats, for instance, do not contribute to the car's driving. But the thesis of the phenomenon-dependence of decomposition goes beyond the observation that not all parts of a system are working parts. Part decompositions generated by different phenomena will carve the system in overlapping and incompatible ways.

Here is just one example: the human body has many systems, which are responsible for the various activities of the body. Examples include the cardiovascular system, the respiratory system, the nervous system, the muscular-skeletal system, and the endocrine system. The behavior of these various systems can be explained mechanistically by decomposing them into parts, and showing how the activities and interactions of these parts are organized so as to be responsible for the activities of those systems. The difficulty is that these different systems carve up the body in different ways. It is not the case that there is some unique set of body parts—one way of carving the body—but there are instead parts of these various systems. The cardiovascular system, for

---

[15] In the context of biological explanation, this point was originally made by Kauffman (1970). An early statement of the principle in the New Mechanist literature is in Glennan (1996). Recent discussions of this principle and the relation between perspectivalism and realism in mechanisms can be found in Craver (2009; 2013) and Darden (2008).

instance, has as its major parts the heart, arteries, veins, capillaries, and blood, while the respiratory system has the lungs, diaphragm, windpipe, mouth, etc. These parts overlap spatially and materially. For instance, the arteries, veins, and capillaries of the cardiovascular system interpenetrate the lungs and the muscles involved in the respiratory system.

As an epistemic matter, how we decompose a system into parts depends upon what phenomenon we seek to explain, so there is an essential pragmatic element in the process by which we decompose systems. But this fact should not lead us to an anti-realist conclusion that parts only exist in the eye of the beholder. Bodies really do have hearts, veins, and arteries, as well as lungs, bronchi, and alveoli—even if these parts spatially and materially overlap. Once the phenomenon is chosen, the scientist can search for the joints in the system responsible for that phenomenon.[16]

Very often, the same object, with the same boundaries, will be an acting part of many mechanisms. Consider the variety of activities that a single animal can engage in—hunting, feeding, defecating, hiding, mating, building, and fighting. In engaging in these activities the same entity is contributing to several distinct mechanisms. But ultimately, the boundaries of the individual will be determined by the specific mechanism, and sometimes those boundaries do not coincide. Biology, for instance, provides numerous examples where the individuals that are parts of evolutionary mechanisms are different from the individuals that engage in ecological interactions. Drones in bee colonies, for instance, are physiologically distinct entities that contribute to the maintenance of the colony. But drones are not reproducers and hence are not individuals within the context of the mechanism of evolution by natural selection.[17] To summarize, then, parts can be real (in the sense of mind-independent) parts even though the decomposition of a whole system into parts is phenomenon-dependent and the parts in different decompositions will overlap.

Let us turn next to the criteria that can be used to identify boundaries. The *locus classicus* of discussions of object/system boundaries is Herbert Simon's account of near-decomposability (Simon 1996; see also Wimsatt 1972). As Simon puts it, a system is nearly decomposable into subunits when the strength or quantity of interactions between those subunits is substantially less than the strength or quantity of interactions among parts of those subunits. For instance, a molecule is nearly decomposable into

[16] To say that parts and their boundaries are objective but potentially overlapping in this sense is not to take a position on metaphysical debates on the nature of part-whole relations, for instance those found in Hüttemann (2010); Schaffer (2009b); Sider (2007); and Winther (2009). One could accept the mind-independence of part boundaries without thinking that parts are prior to or independent of wholes, and one could accept that parts exist at higher levels of organization, while still adhering to a microphysicalist metaphysics, i.e., one that suggests that there is a fundamental level of metaphysical objects or atoms upon which all other things metaphysically depend.

[17] The relation between evolutionary (or Darwinian) and physiological and ecological individuals has been thoughtfully analyzed by Peter Godfrey-Smith (2009a). Godfrey-Smith's analysis of Darwinian individuals starts with the observation that you cannot define a Darwinian individual except in the context of a Darwinian population. This seems to me to be a specific application of the principle of the phenomenon-dependence of decomposition.

atoms because the forces holding together a molecule are substantially less strong than those holding together atoms within the molecule. In another example, Simon asks his readers to imagine a building consisting of a number of thermally well-insulated rooms, each of which in turn contains a number of cubicles. In such a case, the building is nearly decomposable into the rooms, because the thermal conductivity within rooms is much less than the thermal conductivity between rooms.

Simon's characterization cannot be the whole story. For one thing, the account just described presupposes the basic parts, and offers principles for aggregating these parts into larger units. More importantly, Simon's examples might be taken to suggest that there is only one kind of interaction between parts—but typically interactions between parts take place on a number of dimensions (corresponding to different kinds of activities). This thought leads naturally to an elaboration of Simon's criterion that we can, following Grush (2003), call a bandwidth criterion. The idea here is that boundaries between components should be drawn by looking at not just some overall strength of interaction, but also the kinds of interaction. Only limited kinds of interaction will occur at interfaces between components. Computer components provide obvious examples. There is a low bandwidth interface connecting my mouse to the computer. The signal traveling between the mouse and the computer specifies a direction and rate of speed of the mouse, as well as the state of its buttons. Everything else happening within and to the mouse has no impact on the computer.

While many objects have clear spatial boundaries, decompositions into parts are not in the first instance about drawing spatial boundaries, but about drawing causal boundaries. I shall follow Haugeland (2000) in calling these causal boundaries "interfaces."[18] As Haugeland and Simon have both recognized, these interfaces or causal boundaries are not always marked by stable discontinuities in spatial regions. What is essential to a boundary is that it defines the place at which the object may interact with other objects, as well as the kinds of interactions in which the object can engage. Place here is not always to be understood literally as physical location, but more broadly as the method by which it is possible to interact with the object.

Socially defined entities provide good examples of objects that can have clear interfaces without having stable spatial boundaries. Take committees, for example. Committees are system mechanisms that are parts of larger system mechanisms. For instance, a hiring committee is a collection of parts—the members—who engage in activities and interactions that are so organized as to produce certain behaviors, like winnowing large piles of CVs into a set of finalists for campus visits. The committee's activities feed into a larger mechanism, which involves further parts, like the department, the dean, the human resources department, and so on.

---

[18] I originally used the term "interface" in Glennan (1992), having borrowed it from the language of modular and object-oriented programming. In Chapter 4, I discuss the applications of this conception of object boundaries to facilitate abstraction in our representations of mechanisms.

Though the boundaries of social mechanisms are not defined spatially, this is not to say that the committees are not corporeal things.[19] The committee consists of its members, its members have physical locations, and hence the committee has a physical location, namely the union of regions containing its members. But the operations of the committee are not constrained or organized by their physical location. The members of the committee change their physical locations with respect to each other all the time, and with that the spatial boundaries of the mechanism change as well; but, within reason, such changes do not impact the committee's behavior.

The boundaries of objects that are mechanism parts are instead defined by the interface at which they interact with other objects in the mechanisms of which they are part. Suppose that the hiring committee in question is operating under strict rules of confidentiality, so that all communication to and from the committee goes through the chair. If the committee is so organized, then the chair essentially becomes the boundary of the committee by defining and limiting the interactions between the committee and the rest of the world. The bandwidth of interaction is narrow. In other types of committees, all members of the committee may be "on the boundary" in the sense that interactions between those on the committee and those outside the committee may affect the operation of the committee. Here the bandwidth is wider.

But whether the bandwidth at the interface is narrow or wide, there are certain kinds of activities and interactions that take place only among members of the committee. These activities and interactions "inside the committee" are responsible for the behavior of the committee at its interfaces, i.e., in its communications to other entities and individuals in the organization. This example suggests that the inside/outside boundary cannot always be measured on some metric of "strength of interaction," but rather must appeal to different kinds of (organized) activities and interactions that take place on either side of the boundary.[20]

Reflection on academic committee structures and functions provides further evidence of the way in which systems decompose in different ways with respect to different phenomena for which they are responsible. As any faculty member knows, there are many committees, and typically a single faculty member may be part of different committees, so decompositions into parts relevant to the activities of various committees will overlap. This does not make the committees and their boundaries unreal—for with respect to each of the committees' activities (tenure and promotion, grant evaluation, student affairs...), these committees will have clear interfaces that will define interactions between the committee and larger mechanisms.

---

[19] For this reason, I think it is misleading for Haugeland to speak of "incorporeal interfaces."

[20] While Simon used social organizations as examples of systems where causal and spatial organization are independent of each other, this independence is by no means confined to social mechanisms. For instance, it is the selective affinities of ligands to bind with different biomolecules rather than the spatial arrangement of molecules within inter- or intra-cellular solutions that is constitutive of the causal organization of biochemical processes. See the discussion of affinitive mechanisms in Chapter 5.

Even in cases where spatial boundaries are clear, causal boundaries are what define interfaces. Take cell membranes, for instance. Although cell membranes may constitute a clear spatial boundary of the cell, it is not the spatial characteristics that typically define the boundary, but rather, like the committee chair, the membrane's capacity to control and limit interactions between entities inside and outside the cell. Cell membranes are selectively permeable, allowing certain kinds of ions and molecules to pass through at controlled rates and times. The cell membrane thus maintains conditions within the cell that allow the various intracellular mechanisms to operate.

A related way to characterize carving principles is to say that carving should break up a system along boundaries that will identify stable objects in the sense discussed in Section 2.3. Ordinary middle-sized objects typically maintain their shape, color, and mass, along with many other properties that we typically characterize dispositionally or functionally. My toaster reliably toasts, my dog regularly eats, drinks, and barks at squirrels, and the roots of my houseplants regularly absorb water and nutrients. A part exhibits stability both by maintaining a cluster of properties over time and by maintaining these properties in the face of perturbations. Thus a chair maintains its shape over time, and it also maintains its shape in the face of perturbations like lifting the chair and taking it to another room. In this way, the concept of stability is connected with another commonly suggested principle for identifying real parts—namely manipulability. One should divide up a system into objects in such a way that it is possible to intervene on those objects to alter the behavior of a system.

Arguably the stability criterion (and with it the idea that parts are objects) is the most important principle for boundary drawing. As we have seen, parts of a system may interact with their environment with varying levels of intensity and variety; while narrow bandwidth makes it easier to identify boundaries (and is characteristic of modular systems), so long as you have stable properties and clear interfaces, you can have real boundaries, even if they are of high bandwidth.[21]

Given what has been said about entity and system boundaries, we may more briefly turn to the problem of drawing activity and process boundaries. Many of the same principles apply. As with entities, we will get different and incompatible decompositions of mechanistic processes into activities depending upon what phenomena we are seeking to explain. Unlike entity/system boundaries, activity/process boundaries are in two distinct dimensions. First, there is the temporal dimension of duration— when does an activity or process start up and finish? Second, there is a dimension that determines the boundary between the process or activity and its context or background

---

[21] I am arguing for a similar position to the one suggested by Grush, when he argues that distinct parts or components can have wide bandwidth interfaces. Grush suggests that the real criterion is what he calls the plug criterion: "components are entities that can be plugged into, or unplugged from, other components and/or the system at large" (Grush 2003, 79–80). Pluggability, I should note, is a particular species of manipulability (specialized to control systems), and while I would agree that pluggability is a better marker of a genuine component than bandwidth, I would suggest that what ultimately makes something a genuine pluggable component is a set of stable powers intrinsic to that component.

conditions; I will call this the breadth of the process. As with boundary questions about systems, boundary questions about activities and processes involve both the drawing of an external boundary between the process as a whole and the articulation of the parts, here stages, within the process.

As a first example of how temporal boundaries are drawn, consider again the hiring committee. The main activity of the hiring committee (i.e., the phenomenon for which it is responsible) is recommending a candidate; this activity starts with the review of candidate dossiers and terminates with a recommendation to the dean. The activity may be broken up into sub-activities: for instance, the committee may choose semifinalists, conduct screening interviews, invite finalists to campus and so forth. The reason one sets the boundaries of the committee's recommending activity in this way is that these are the boundaries within the embedding mechanism. The dossiers are fed to the committee by other entities, and it is the committee's final recommendation that triggers the activities of the dean. There are no doubt important idealizations that are involved in drawing these boundaries; for instance, the dean may interact in various ways with committee members prior to receiving the recommendation. But the boundaries are not arbitrarily chosen. There are clear and generally low-bandwidth interfaces that mark the beginning and ending of the various activities.

A single activity may be embedded in many larger processes, and how one should carve up the activity—to set its beginning and endpoints (and intermediate stages)—will depend on which of these larger processes one is seeking to explain. A vivid example is human sexual intercourse. When does it start and end? Does it begin with a comment over dinner, a kiss, hopping into bed, or penetration? Does it end with fertilization, withdrawal, with orgasms, with snuggling, or sleep? The variety of concrete activities that fall under the term "sex" are embedded with a variety of mechanisms—physiological, reproductive, psychological, social, economic, political, and evolutionary. How we characterize the boundaries of this activity will be determined by which of these mechanisms we are seeking to explain. A psychologist interested in understanding the conditions under which humans choose when and with whom to engage in sex must take a wider view than a physiologist seeking to understand the physiology of orgasm or a reproductive physician seeking to understand the conditions of successful fertilization.

While the duration question concerns when activities start and stop, the breadth question concerns how many entities are involved in a single activity and how spread out these entities are through space. To illustrate the issue, consider the activity of dancing. Suppose Franz and Sisi are at the ball and start to dance. Is this an activity they engage in singly, as a pair, or as a larger group? Perhaps the most natural answer is that it depends upon the dance. Some dances are single dances, others pair dances, and still others group dances. For instance, the quadrille is an eighteenth- and nineteenth-century dance involving four couples. The waltz on the other hand is a pairs dance. So, if Franz and Sisi are waltzing, their waltzing is one activity, while the other couples' waltzing are different activities; alternatively, if they are doing the quadrille, they are part of an activity involving eight dancers. As always, these intuitive judgments are grounded in the phenomena one seeks to explain. For instance, if one is seeking to understand the patterns of movement across the dance floor, in the quadrille these are produced by organized interactions of eight dancers; in the case of the waltz, the pattern of each pair is more or less independent of the other. And again, the "more or less" reflects the idealizations involved in separating one activity from another. It is evident that Franz and Sisi's dance movements are constrained by the locations of other dancers on the floor. But we might say, following Simon, that the interactions of waltzers are nearly decomposable into couples, while the interactions of those dancing the quadrille are not. Note also that this characterization of waltzing as a pairs activity follows from focusing on the particular phenomenon of motion on the dance floor as one's explanatory target. One can choose to focus on other phenomena as well; for instance, if our explanatory target is physiological changes in Sisi—changes in her heart rate, breathing, etc.—then waltzing can be construed as a solo activity, and Franz moves into the background conditions.

Finally, let me turn to mechanism boundaries, the criteria for which are related to but not identical to object/system and process boundaries. As we noted above, objects/systems are persistent and stable entities. They have mechanisms, but they cannot be identified with mechanisms, which involve activities at a given point in time. To put the distinction in the language of Craver diagrams, object/system boundary questions concern which Xs fall within the boundaries of a system S, while system mechanism boundaries concern which working parts, which Xs Φ-ing, contribute to the activity of the mechanism as a whole, S ψ-ing. For instance, a cell is a system S with boundaries, and many entities reside and activities go on within the cell. If we take one such activity, say synthesizing a particular protein, we can ask which entities and activities within the cell actually contribute to the production of the protein.

Craver has called this latter problem the problem of constitutive relevance. His use of that term is deliberately evocative of the concept of causal relevance. To say of an event that it is causally relevant to an effect, is to say that it *made a difference* to the effect. Had the event not occurred or had it had different properties, the effect would have been different. The difference between causal relevance and constitutive relevance is that causal relevance concerns how causes make a difference to their effect, while constitutive relevance concerns how the activities of parts make a difference to the activities of wholes.

Craver's preferred account of this relation involves what he calls the mutual manipulability criterion. The criterion suggests that if you manipulate a working part of the mechanism it will change the behavior of the whole mechanism, and if you manipulate the whole mechanism, say by changing its "inputs" or environment, it will change the behavior of its parts. The theory takes as its starting point Woodward's (2003) manipulability-based account of causation. The difference is that with causes and effects, manipulation occurs only in one direction, from causes to effects, whereas in the constitution relation, manipulation occurs in both directions, bottom-up, from parts to wholes, and top-down, from wholes to parts.

Craver's account has been the subject of much discussion, mostly focused on whether you can use what appears to be causal criteria for constitutive relationships (Leuridan 2012; Baumgartner and Gebharter 2016). In my view, the chief problem with the mutual manipulability account is that it mistakes an epistemic criterion for an ontological one. Craver's criterion offers an essential method for distinguishing working parts from non-working ones and it shows how that method is related to, but distinct from, methods for identifying causal relevance. Mutual manipulability is thus part of the boundary-drawing toolkit, but it does not provide an answer to the ontological question of what the mechanistic constitution relation is.[22]

Let me briefly recapitulate the results of our discussion of boundaries. The fundamental point is that boundary drawing—whether spatial boundaries between parts of mechanisms or between a mechanism and its environment, or temporal boundaries between the start and endpoints of an activity or mechanical process—has an ineliminable perspectival element. But the perspectives from which these boundaries are drawn are not arbitrary or unconstrained. The perspective is given by identifying some phenomenon. This phenomenon is a real and mind-independent feature of the world, and there are real and mind-independent boundaries to be found in the entities and activities that constitute the mechanism responsible for that phenomenon. In this section, we have identified some of the features that characterize these boundaries.

2.6 Mechanisms, Regularities, and Laws

Advocates of New Mechanism have often framed their accounts as alternatives to the logical empiricist tradition (Craver and Tabery 2015), and this is nowhere clearer than in the New Mechanist analysis of laws of nature. New Mechanists have argued that mechanistic explanation is to be preferred to Hempel's covering law models of explanation (Glennan 2002a), that mechanistic approaches provide a superior account of inter-level relations to those founded on law-based models of inter-level reduction (Craver 2007; Glennan 2010a), and I at least have argued for the possibility that a mechanistic account of causation could replace ones that implicitly or explicitly appeal to laws (Glennan 1996; 2011). While New Mechanists have often downplayed the role of laws in science (Machamer, Darden, and Craver 2000), mechanistic approaches do

---

[22] Couch (2011) and Harbecke (2010) have offered regularity-based accounts of constitutive relevance that they think avoid problems in Craver's account. The idea, again, relies on a concept borrowed from the causal literature. Borrowing Mackie's idea of an INUS condition (that a cause is an insufficient but necessary part of an unnecessary but sufficient condition), Couch, for instance, defines a constitutively relevant part as "an insufficient but nonredundant part of an unnecessary but sufficient mechanism that serves as the realization of some capacity" (Couch 2011, 384). Although Couch would argue otherwise, I think that this again mistakes epistemic for ontological considerations. If this sort of INUS condition is understood as describing a regularity, it makes the question of whether or not the mechanism is a working part of a mechanism depend upon facts that are extrinsic to the mechanism. This runs counter to the view of mechanisms and causation that I am developing here. If the singularist approach advocated in this book is correct, regularities provide evidence for working parthood, but are not what make parts working parts.

not need to deny that there are laws. Rather, the point is to argue that a great variety of scientific generalizations, some of which we call laws, are in fact explained by mechanisms.

The basic New Mechanist account of laws and regularities is pithily captured in Holly Andersen's formulation that "regularities are what laws describe and what mechanisms explain" (Andersen 2011, 325). The idea is that many generalizations that we have or might call laws are simply descriptions of the behavior of mechanisms; that is, the phenomena for which the mechanism is responsible is a lawful regularity, and the mechanism that is responsible for that regularity explains the law.[23]

A classic example of such laws are Mendel's laws. The regularity described by Mendel's law of independent assortment (the second law) is this: Given any two loci in a diploid organism, the probability of transmitting a particular allele at the first locus is independent of what occurs at the second locus. So, for instance, in an organism with genotype AaBb, the probability of transmitting alleles A and B are independent, so $P(AB) = P(A)P(B)$. Combined with the segregation law, which states the probability of transmitting each allele from a given locus is equal, we get the following regularity: $P(AB) = P(Ab) = P(aB) = p(ab) = .25$.

The regularities described by Mendel's laws depend upon the mechanism of gamete formation. Genes are located on paired chromosomes, with the two alleles at a single locus coming from corresponding spots on homologous chromosomes. In gamete formation, half of the genetic material is copied into each gamete. If the loci are on different chromosomes, the probabilities of a given gamete receiving genetic material from one or the other homologs is independent. Even if the loci are on the same chromosome, genetic recombination will lead to independent segregation so long as the loci do not lie too close together on the chromosome. But Mendel's laws are not strict laws. The regularities they describe have many exceptions. Linked genes do not assort independently, violating the second law, and (more rarely) segregation distortion mechanisms will make one allele at a locus more likely to be transmitted than another, violating the first law.

The lawful regularities that mechanisms can be said to explain are never strict laws. They are not of unrestricted scope because they depend upon the particulars of which the mechanism is made. This accounts for the ceteris paribus character of the lawful generalizations. If a lawful regularity obtains because it is the product of a mechanism,

---

[23] It is important to keep track of a persistent ambiguity in the philosophical literature on laws. Laws are sometimes taken to be things in nature—either regularities of certain kinds (on Humean analyses) or some sort of power or dependency in the world that accounts for regularities. Alternatively, laws are taken to be generalizations, which are *statements* that describe these regularities. I am, like Andersen, taking the latter approach. That is why laws describe regularities that are the behaviors of mechanisms. The alternative view distinguishes between laws (in the world) and law statements that refer to those laws. If for instance, laws are relations between universals (Armstrong 1983), they clearly are not statements. Both scientists and philosophers move back and forth between these senses, e.g., between laws governing phenomena (where laws clearly are not statements) to theories containing laws (where laws clearly are statements). While I cannot see a definitive reason to privilege one usage, I concur with Craver and Kaiser (2013) that failure to attend to this distinction can lead to mischief.

that regularity will only hold so long as the mechanism does not break down. Mendel's laws only hold true in the absence of segregation distorters and for non-linked loci. Lawful generalizations describing chemical reactions in a solution will depend upon background conditions like the temperature and pH of the solution. Cars will only start when they have gas in them.

The New Mechanist account of laws is a deflationary one, as there are many laws and they are of restricted scope. Robert Cummins has put it nicely in his discussion of laws in the special sciences:

Laws of psychology and geology are laws in situ, that is, laws that hold of a special kind of system because of its peculiar constitution and organization. The special sciences do not yield general laws of nature, but rather laws governing the special sorts of systems that are the proper objects of study. Laws in situ specify effects—regular behavioral patterns characteristic of a specific kind of mechanism.   (Cummins 2000, 121)

While framed particularly in terms of sciences like psychology and geology, the point applies equally to many generalizations in physics. For instance, generalizations about the behavior of materials (diamonds, springs, fluids) depend essentially on the constitution and organization of the materials of which they are made.

Cases like Mendel's laws illustrate the New Mechanist account of laws as mechanism-dependent, but the relationship between laws, regularities, and mechanisms is not always of the sort characterized by this paradigm case. Consideration both of the characterization of minimal mechanism and of exemplars of laws, regularities, and mechanisms in the sciences suggest that mechanisms need not always be regular, that the phenomena for which mechanisms are responsible need not always be described by laws, and that some generalizations commonly called laws describe patterns or regularities that are not mechanism-dependent. Let us take these three points in turn.

First, natural phenomena and the mechanisms responsible for them can be regular or irregular to differing degrees and respects. The paradigm cases of mechanisms, like the mechanisms responsible for Mendel's laws, describe what Andersen (2012), following MDC, calls "always or for the most part" regularities. These are regularities in a mechanism's behavior that will always occur unless there are breakdowns, or background conditions are violated. But many mechanisms do not work "always or for the most part." Consider mechanisms responsible for transmission of infections. Many viruses, bacteria, or parasites will infect a host with only a small probability, even if there is a determinate and common mechanism by which they do so when they do.

Beyond the question of whether and to what extent the phenomena for which mechanisms are responsible must be regular, we must also ask whether and to what extent the mechanisms responsible for producing these phenomena are regular. These varieties of (ir)regularity are not the same (Glennan 1997a). It is possible for a mechanism's phenomenon to be in various respects irregular, while the behavior and organization of its constituents are regular, or conversely for irregular behavior and organization of a mechanism's constituents to be responsible for regular behaviors in the mechanism as a whole. For instance, roulette wheels and deterministic random number generators can produce phenomena that are in obvious respects irregular, while irregular and uncoordinated interactions of components of large systems (be they molecules, pedestrians, or something else) can give rise to highly regular aggregate behavior (see Section 5.4).

Second, not all the phenomena for which mechanisms are responsible are described by laws. One reason for this is that there can be regular phenomena for which mechanisms are responsible, but that one would not describe as lawful. A case in point is common causes. There is, famously, a regular association between barometers dropping and storms coming, but it is not causal and few would claim that that association counts as a law. There are, however, mechanisms accounting for both the barometer drop and the storm coming, and since they are triggered by the same cause, their association is an always or for the most part regularity. Another kind of phenomena that is not law-like are one-off phenomena, like chance meetings at the park or a collision of the earth with an asteroid. Such phenomena are the result of ephemeral mechanisms (Glennan 2010b; 2014).

Third, some regularities described by laws do not depend upon mechanisms at all. Most obviously, laws that describe general constraints on physical systems, like the laws of thermodynamics and conservation laws, cannot be plausibly understood as describing the behavior of particular mechanisms. Also, it may turn out that there are some basic laws relating physical quantities or events that are not dependent upon mechanisms. For instance, this would be the case for the law of universal gravitation if Newtonian gravitational attraction were a basic property of matter.[24]

One of the reasons regularities are important is that they provide grounds for drawing boundaries for entities, activities, and mechanisms. In Section 2.3, I emphasized that systems (and system mechanisms) are characterized by their repeatable behavior and stable dispositions. These are regularities. Similarly, the principles for carving boundaries discussed in Section 2.4 suggest that boundaries between parts, and between mechanisms and their environment, should be drawn so as to pick out stable objects. Here again, there is regularity. Because of the epistemic importance of regularity, Andersen has argued that regularity is essential to a scientifically useful conception of mechanism. She grants that minimal mechanisms may play a role in metaphysical issues concerning causation, but thinks that the lack of regularity in them makes them not central to the scientific enterprise (Andersen 2011; 2014a; Baetu 2013; Levy 2013).

In response to Andersen, I would make a couple of points. First, while the sciences are concerned with regular and repeatable phenomena, scientists, both natural and

---

[24] I hope I may be forgiven for an example from a dated physical theory. Classical force laws provide good illustrations of the notion of mechanically inexplicable laws, even if the classical mechanics on which they are based has been superseded. While the interpretation of some basic principles of modern physics (e.g., the Einstein field equations or the Schrödinger equation) is more complex, I take it that it is not plausible to think of such principles as mechanism-dependent. For a summary of the prospects of mechanisms in physics, see Kuhlmann (2018).

social, are also interested in understanding singular phenomena. Historical explanation, both human and natural, is principally concerned with accounting for the causes of singular events. Second, the business of identifying the ephemeral mechanisms responsible for one-off occurrences does in fact depend in important ways on regularities. So, for instance, if a toddler accidentally shoots herself with a handgun that was carelessly left about, the tragic situation, while it involves a lot of non-repeatable accidents about where and why the toddler and the gun were where they were, is explicable in terms of regularities in the dispositions of toddlers (like their curiosity) and of guns (like their firing bullets when triggers are pulled).[25]

2.7 New Mechanical Ontology

I began this book with the assertion that the New Mechanical Philosophy, at least as I understand it, is both a philosophy of nature and a philosophy of science, and that its central thesis regarding nature is that practically the whole sweep of natural phenomena as described by the physical, life, and social sciences, are the product of mechanisms. Now, with a characterization of what (minimal) mechanisms are somewhat fleshed out, I would like to summarize the ontological picture that emerges, and to say something about how the categories I have used in describing mechanisms relate to some more well-known denizens of the metaphysician's zoo.[26]

Metaphysicians of course disagree sharply on questions of ontology, so a new mechanical ontology cannot readily be compared to a "standard view." Even confining myself to the contemporary literature, I cannot hope to survey the whole variety of ontologies on offer. My goal is more modest—to say something about how some of the ontological and metaphysical concepts and relationships that have been of interest to contemporary (and sometimes classical) philosophers relate to the ones I have used to characterize mechanisms.

---

[25] In her argument for a mechanistic account of singular causal chains, Andersen (2012) makes two claims. The first is that without regularity you cannot identify which entities and activities are working parts, i.e., causally relevant to the mechanism. I would argue, however, that this is just what historians do when they construct narratives (which are essentially descriptions of singular causal chains). The work is fallible, and it is based in large part on knowledge both of proximity (what could interact with what) and the regular dispositions of the potential actors. Andersen's second point is that in describing singular causal chains responsible for one-off events, there is no objective way to determine a place in the chain that represents where the mechanism starts. Here I would agree: where we choose to start a narrative describing the causes of one-off events depends upon our goals and interests. But given a pragmatically chosen starting point, there is an objective answer to the question of how that starting point is productively connected to the one-off event.

[26] Ontology is often construed as being a subfield of metaphysics, involving basic questions of the categories of things that make up the world, while metaphysics is taken to include a wider variety of questions—from the nature of causation to the structure of time to freedom of the will. I will, however, use the terms loosely. My main concern is to distinguish ontological and metaphysical questions from epistemological, methodological, and semantic questions—or, as I have put it more colloquially, questions about nature from questions about science.

Different metaphysicians use different tools, but it is common, both historically and in contemporary discussions, to characterize ontology in terms of the categories of substance, property, relation, and law.[27] Substances are things or objects—particulars located somewhere in space and time. They include things like electrons, mountains, and brains. Properties are instantiated in particular substances, and it is the properties that particular substances have that make the substances the kinds of substances they are. There are also various kinds of relations that may obtain between substances— spatial and temporal relations as well as causal. Many metaphysicians believe that causal relations between substances depend upon the properties of those substances in combination with laws of nature.

One important issue for metaphysics concerns the relationship between simples and composites. For instance, mountains and brains are composites—things made of other things—while electrons are (we think) elementary particles that do not themselves have structure or parts. While few doubt the existence of composites, there are disagreements over their status. For instance, are composites like brains not true substances? Relatedly, are the properties we attribute to composites truly properties, or only, as John Heil (2012) puts it, "properties by courtesy"? And, if laws are relationships between properties or universals, are the only true laws fundamental laws that relate the few "sparse" properties of basic substances?

If one wants to countenance non-fundamental substances and properties, one needs an account of the relationship between the fundamental ones and non-fundamental ones. Particular substances (at least the non-fundamental ones) are thought to have parts, though exactly what it takes to be or have a part is controversial. For properties, the most common move is to appeal to supervenience, where supervenience is understood as a relationship between two sets of properties—for instance, physical properties and mental properties. The idea is that if two things differ with respect to the supervening properties, they must also differ with respect to the base properties. Alternatively, the relation between properties might be one of realization, where the more basic properties realize the less basic ones. Realization approaches typically take the higher-level properties to be functional properties, individuated by causal powers, while the realizing properties are what give things with those higher-level properties those causal powers. Another recently popular candidate is the so-called grounding relation, which is typically taken to be some "hyper-modal" dependency relation (Schaffer 2009a; Rosen 2010; Fine 2012). With each of these approaches, we get different accounts of what to make of laws involving higher-level properties.

Let us then consider how these metaphysical categories and relationships map on to mechanisms. We begin with the New Mechanists' entities. As I already indicated in Section 2.1, the category of things New Mechanists have referred to as entities coincides with common conceptions of substance, at least those conceptions that countenance

---

[27] I will ignore for now the currently popular notion of structure (Ladyman 2014).

compounds as true substances. Entities are particulars in space and time that persist and are the subject of change. Not much is new here.

Activities on the other hand seem to be a novel ontological category. They cannot naturally be reduced to properties of or relations between entities. Given the apparent ubiquity of activities and interactions in the natural world, it may be surprising that this category does not typically find its way into debates in contemporary metaphysics. I have a hunch that the reason traditional metaphysical categories fail to do justice to activities is that these categories are artifacts of our infatuation with our systems of logic. Any student of elementary predicate logic will see that it is a language of sub-stances, properties, and relations.[28] It is not a language with real verbs. The central logical operation is predication, which is the attribution of a property to a substance, or the attribution of a relation to an (ordered) set of substances. The entities within a domain are described as having properties or standing in relations, but not as doing things. We can say of Fido that he is a dog "Df," but we cannot so easily say that Fido walks. We can and do jury-rig, as when we attribute the property of walking to Fido "Wf." Formally this may work, but it reduces doing to having; it takes the activity out of activities.

We see a similar issue in our characterization of causality. On the New Mechanist account, activities are causes, in the sense that it is entities engaging in activities that are the producers of change. Contrast this with standard moves in metaphysical discussions of causation. Here one begins with the questions, what is the nature of the causal relation, and what are its relata? Typically causation is conceived as a two-place relation Cxy over a domain of things—most often events.[29] Formally the causal rela-tion looks quite similar to non-causal relations like is-taller-than or is-older-than. This observation should not be taken to imply that philosophers do not recognize the differences between causal and non-causal relations; but it does remind us that the language of relations is a static language, and that the forms logic provides to charac-terize the metaphysics of causation can be misleading.

And what of laws? Again, logical categories shape accounts of metaphysical relations. The traditional empiricist account treats laws as a species of universal generalization, variations on the form of "All Fs are Gs," or "$\forall x(Fx \rightarrow Gx)$." To recall some venerable examples, it is taken to be a law that all emeralds are green, that all copper conducts electricity, or that all uranium spheres are less than a mile in diameter. The traditional challenge is to distinguish these lawful generalizations from accidental generalizations— like the claim that all the coins in my pocket are silver. Critics of the empiricist trad-ition like David Armstrong take a somewhat different route, suggesting that the right way to think about laws is as relations between properties or universals, so all Fs are Gs is interpreted not as a universal generalization about entities having F and G, but as

---

[28] I am not a student of Aristotle's logic or metaphysics, but my sense is that my remarks about the meta-physical biases that we get from our logic apply to (and to some degree come from) Aristotle, though of course Aristotle was deeply concerned with the nature of change.

[29] This is certainly not the only way the causal relation is conceived in the metaphysics literature. For a succinct description of alternatives see Schaffer (2014).

a singular claim about a relation between F and G themselves. Armstrong (1983) symbolizes the relation N(F,G), where N is a non-logical second-order necessitation relation. My point is not to take a position on any of these accounts, but simply to observe that these two most familiar ways of thinking of laws (often called the Mill-Ramsey-Lewis and Armstrong-Dretske-Tooley approaches) both adopt the static language of substances, properties, and relations. Activities are hard to discern in either of these accounts.

The dynamic character of activities and laws can easily be obscured by almost any act of scientific representation. When one puts pen to paper, using static inscriptions as representations of dynamic and temporally extended processes, something will always be lost. Nonetheless, philosophical representations seem especially weak on this score, and more standard scientific approaches to characterizing activities and causal relations seem markedly better. Consider for example the role of differential equations for representing the dynamics of the systems they study. When scientists characterize patterns as causal laws, or when they represent mechanisms in the world, differential equations are very often the means by which they are represented. They are used across the gamut of sciences, from physics and chemistry to psychology and economics. This is no surprise, as the calculus was invented and developed by Leibniz, Newton, and their successors for the specific purpose of characterizing change. Why, one might wonder, do metaphysicians read their ontological categories off of predicate logic and its higher-order and modal extensions, rather than asking what ontology falls out of the representations of nature that science offers us?[30]

Though I doubt that the language of properties and relations is sufficient to capture the concepts of activity and interaction that are so central to understanding mechanisms, this is not to deny the existence or importance of properties for an account of mechanisms. MDC explain:

Activities usually require that entities have specific types of properties. The neurotransmitter and receptor, two entities, bind, an activity, by virtue of their structural properties and charge distributions.   (Machamer, Darden, and Craver 2000, 3)

In Section 2.4 I have argued that the properties of compound entities (systems) depend upon mechanisms. In characterizing what phenomena mechanisms can be responsible for, I suggested that we can speak not just of the mechanism responsible for entities acting (S ψ-ing), but also of mechanisms responsible for capacities of entities. So we can speak of the mechanism responsible for the cow's digesting the grass (when she's actually digesting) but also of the cow's digestive system as a mechanism that gives the cow the capacity to digest when she eats. Capacities, as I understand it, are just dispositional properties of systems. Appropriately coupled to other processes and systems

---

[30] There is of course a strong and diverse tradition within the philosophy of science community that takes this latter course. Debates, for instance, about the foundations of quantum mechanics or the interpretation of probability typically begin with scientific representations, and, at their best, do not demand as an a priori constraint that their ontological conclusions cohere with a substance-property-law ontology.

(its "disposition partners"), those capacities will manifest themselves in the system's activities and its interactions with things in its environment.

What of so-called categorical properties? These too have a place in a mechanistic ontology. A categorical property is an intrinsic feature of an entity, rather than a capacity or disposition of an entity to act in the world. So, for instance, the shape of the key seems to be a categorical property, a way the key is. And while the shape of the key may give the key the capacity to open certain locks, that capacity is not identical to the key's shape. On the New Mechanist account, the relationship between categorical and dispositional properties is analogous to the biologist's familiar distinction between structure and function. When biologists speak of the structure of an entity (a molecule, a cell, an organ, etc.) they are referring to how the parts that make it up are organized— for instance the shape of a molecule or the shape of a finch's beak. These are ways the molecule or beak are, categorical properties. These structural/categorical properties give the molecules or beaks their function (dispositional property), for instance to catalyze a reaction or to break open a seed of a certain shape. In terms of minimal mechanism, these structures, the ways mechanisms are arranged, are captured in the concept of a mechanism's organization.[31]

Let us consider now the relation between more and less basic substances and their properties. The most common answers within contemporary metaphysics appeal to either supervenience relations or realization relations. In their stead, I have argued that New Mechanists appeal to a relation of mechanism-dependence—the capacities displayed by entities and the activities they engage in are mechanism-dependent. I have said some preliminary things about mechanism-dependence already in this chapter, and will have to say more, particularly with respect to the concept of mecha- nism-dependent production, in Chapter 7. For the moment, I will limit myself to a few comments about the relation between mechanism-dependence, supervenience, and realization.

Perhaps the most obvious difference is that both realization and supervenience are typically construed as relations between properties or property families, while mecha- nism-dependence is between a particular phenomenon and a particular mechanism. Consider, for instance, Kim's definition of strong supervenience, which defines a rela- tion between two sets of properties, A and B:

---

[31] Several caveats are in order. First, someone committed to the idea that only simple substances have properties will not countenance structural properties/mechanistic organization as true categorical proper- ties, since these properties essentially have to do with *relations* between entities and activities that make up compounds. Second, while the biological distinction between structure and function is helpful here, it bears repeating that mechanisms need not have functions in the teleological sense. Philosophers are divided about whether dispositional properties reduce to categorical properties, or categorical properties reduce to dispositional properties, or whether they somehow can both exist. I am sympathetic with the argument that fundamental entities (if there be such) will only have dispositional properties—that there is nothing to what they are other than what they do—but for compound entities, which are the New Mechanists' focus, it makes perfect sense to distinguish categorical/structural properties from disposi- tional/functional ones.

A *strongly supervenes* on B just in case, necessarily, for each x and each property F in A, then there is a property G in B such that x has G, and necessarily if any y has G, it has F. (Kim 1984, 165)

Subsuming relationships of dependence between mechanisms and their component entities and activities to relations between supervenient and base properties embodies what I have elsewhere described as property bias (Glennan 2010a). Property bias occurs when one assumes that the properties of an entity depend only upon other properties, rather than understanding that the properties of a compound entity depend upon the organized activities of its particular parts. The dependence of properties on particulars makes the properties of an entity local to that entity.[32]

We find something similar in discussions of realization. Though there are debates about the proper relata for the realization relation, it is often understood as a relation between two properties. The realizing property realizes the realized property if the causal powers that individuate the realized property depend upon the realizing property. In contrast to this, mechanisms are particulars (entities acting) that depend upon other particulars, the component entities and their activities. The properties and causal powers of a mechanism as a whole do not depend upon other properties of the mechanism as a whole, but instead upon the activities of and relations between components. One way of describing the relationship between mechanisms, realization, and supervenience is to say that mechanism-dependence gives an account of how realization works, or why supervenience holds (cf. Endicott 2011).

Gillett's (2003) account of dimensioned realization is more congenial to New Mechanist ontology than traditional accounts. In dimensioned realization, the realizing properties are not properties of the entity exhibiting the realized property, but of that entity's parts. Gillett and Aizawa have argued that dimensioned realization explains the ontological underpinnings of mechanistic explanation (Gillett 2007; Aizawa and Gillett 2009). While I shall not explore the details here, one can think of this work as an attempt to give an account of the nature of mechanism-dependence. It seems to me to be an important step toward clarifying the ontological commitments of New Mechanists. However, in one respect, this more mechanism-friendly account of realization still misses some of what mechanism-dependence is about. Since the account is still very much tied up within the tradition of the metaphysics of properties, it focuses on the synchronic relation of realization, rather than on the dynamic character of mechanisms producing their phenomena over extended periods of time.[33]

---

[32] The idea that properties of entities are mechanism-dependent has affinities with the view that properties are abstract particulars or tropes. Both approaches localize properties and powers in their instances, though the properties and powers of concern to the Mechanists are properties of compounds—properties that do not arise from some bundling of elementary tropes, but from the organized activities and interactions of their constituents.

[33] Aizawa and Gillett do recognize the need to attend to the temporal dimension of mechanisms. In their account they distinguish between realization relations between properties, constitution relations between individuals (entities), and implementation relations between processes (Aizawa and Gillett 2009, 182). My sense is that these dimensions cannot be so neatly disentangled—hence my emphasis on the multi-dimensional notion of mechanism-dependence.

As to laws, the basic New Mechanist account, as outlined in the previous section, is that laws are simply descriptions of the behavior of (regular) mechanisms, so lawful regularities are grounded in and explained by mechanisms. It is true that these mechanism-dependent laws may strike some critics as not being genuine examples of laws; generalizations about how gametes or proteins are constructed may be scientifically useful, but will not get at the metaphysical heart of the matter. The concern is that mechanisms, being composites, cannot ultimately ground nomological dependencies. True laws on the critics' view would be confined to a level of fundamental entities— simple substances, whose interactions are determined by laws, powers, or patterns of counterfactual dependence. But the New Mechanist's ontology is an ontology of compounds. It offers an account of the relations between simpler and more complex entities, activities, and mechanisms. Perhaps, for some, this makes the New Mechanist ontology not an ontology at all. At any rate, it is likely the case that much of what I say is compatible with a number of alternative views on the ontology of the fundamental level, if indeed there is such a level. The question of whether there is a fundamental level, what that level is like, and how such a level (or a lack thereof) could ground compound mechanisms is an important one, but I shall defer discussion until Chapter 7.

Given the centrality of mechanism-dependence to the ontological picture I am arguing for, it would be good to say something more about the vertical dimension of mechanism-dependence that has come to be called mechanistic constitution. In our discussion of boundaries I have already discussed criteria to identify what entities should count as working parts of a mechanism, but here I want to focus on the ontological issue. Supposing we have a multi-level mechanism where some system's activity (S ψ-ing) is constituted by a set of acting and interacting parts (Xs Φ-ing). What, from an ontological point of view, is the nature of the constitution relation?

Consider as a concrete example the mechanism by which muscles contract. Suppose that I contract my bicep. My bicep is an entity engaging in the activity of contracting. How does it do this?[34] The muscle is composed of cells called myocytes, which in turn contain bundles of fibers called myofibrils, each of which are composed of segments of units called sarcomeres. The sarcomeres are themselves compounded of filaments made mostly of two proteins, myosin and actin. Muscle contraction occurs because of the actin filaments sliding along the myosin filaments. The muscle contracts because the myofibrils contract, which occurs because the sarcomeres contract, which happens because the filaments within the sarcomeres slide along each other.

The key question here concerns the relation between constitution and causation. I have said that the muscle contracts because of the sliding of the filaments, and this is true, loosely speaking. But what is happening here is a combination of the horizontal and vertical dimensions of causation and constitution. The filaments slide—this is a causal interaction—but the sliding of the filaments does not *cause* the contraction of

---

[34] This brief description of the anatomy and physiology of muscle is drawn from a recent textbook (Mulroney, Myers, and Netter 2016).

the sarcomeres, because the filaments are *parts* of the sarcomeres. The sliding is not some event in the past, which causes the contraction of the sarcomeres in the future. The sliding and the contracting happen at the same time, because the sliding filaments are what the contracting sarcomeres are made of. In this sense, the contraction of the sarcomeres *just is* the collective sliding of the filaments, the contraction of the myofibrils just is the collective contraction of the sarcomeres, the contraction of the muscle cell just is the collective contraction of the myofibrils, and so on.[35]

This "just is" relationship is the relationship that has been called mechanistic constitution. It is the relation, whatever it is, between acting parts and acting wholes. One may be excused for wondering if this relationship is not just a relation of token identity. When we say the contracting of the muscle just is the contracting of the muscle fibers, is it not the case that the contracting muscle is identical to its contracting fibers?

Traditionally, the reason to make a distinction between constitution and identity has had to do with puzzles about the persistence of objects through time (Wasserman 2015). The problem is that many objects seem to persist even as the matter of which they are made changes. A cloud will gain and lose the water particles that constitute it, while remaining the same cloud. Nowadays humans can lose joints or limbs and have them replaced with prosthetics. They still remain the same people. And of course at the molecular level, all organisms (including humans) are constantly exchanging material with the environment. So you and I may be constituted by the molecules of which we are made, but we are not identical to them. Applying this to the muscle, we can see that it would be wrong to say that the muscle is identical to its fibers, because cells within the muscle tissue live and die, while the muscle persists.

But maybe this concern about the material constitution of entities does not apply to mechanisms. If mechanisms truly are understood to be entities acting, the mechanism persists only for the duration of the action. We are not asking here what constitutes the muscle, but what constitutes the muscle contracting. And when the muscle contracts this one time, can it not be that the contraction just is, in the sense of token identity, the contracting of the fibers? Is it not the case that when we speak of the fibers contracting and the muscle contracting, we are just giving finer- and coarser-grained descriptions of the very same thing? I expect we can speak of identity here (Glennan 2010a) but, given the manifold confusions that arise when speaking of identity of objects over time, one might prefer, as Povich and Craver (2017) suggest, to call this "just is" relation "exhaustive constitution" and leave it at that.

Let me conclude my account of New Mechanist ontology with a discussion of its implications for one version of the monism/pluralism debate. The version I am interested in is the one concerned with the relationship between concrete objects or substances in the universe and their parts (Schaffer 2009b). The monistic position is that the whole (the cosmos) is prior to the parts, with the parts just being aspects of the

---

[35] This account of the relation between causation and constitution is essentially the same as that suggested by Craver and Bechtel (2007). It will be discussed further in Section 7.9.

whole, while the pluralist position is that the whole cosmos is made of parts, with the whole then depending on those parts. Because New Mechanists emphasize the ways in which the organized activities and interaction of parts explain the behavior of wholes, it might seem that New Mechanist ontology is committed to a pluralist (and reductionist) view.

While it is true that New Mechanist ontology requires that parts be real things that explain the behavior as wholes, I want to forestall a couple of possible confusions about the consequences of this position. First, while there are plausible interpretations of the ontology of quantum mechanics that support a kind of holism about the cosmos (Schaffer 2009b), there appear to be ways to square this quantum holism with a mechanistic ontology of real parts (Kuhlmann and Glennan 2014). Second, a consideration of the varied kinds of relations between mechanisms and their parts will show that there are different senses in which wholes can be prior to or subsequent to their parts, and that a mechanistic ontology can allow that in certain kinds of mechanisms there are empirically significant senses in which parts depend upon wholes.

The most obvious sorts of priority which can be given an empirical meaning seem to me to be causal and historical priority, and these questions must be judged case by case. In the case of a car engine, for instance, there is a non-problematic sense in which the parts precede the whole. The parts of the car engine (belts, pistons, plugs) were fabricated beforehand and assembled later. But in the case of living systems, parts (organs, tissues, etc.) are grown by and within the whole, not brought from another factory. Another question we might ask is the extent to which a part can remain the part that it is in the absence of the whole. Nuts and bolts can be taken out of the machine of which they are part and still remain nuts and bolts, but it's hard to put components of organisms on the shelf in the same way.

These examples might be taken to suggest a fundamental divide between living and non-living systems, with living systems having a holistic character absent in the non-living ones. But while I think that living systems are paradigms of systems in which parts depend upon the wholes of which they are a part, the situation is by no means unique. All objects that can be parts of mechanisms maintain their stability only within some range of mechanistic and background conditions: an iceberg is only an iceberg (or for that matter, the bolt is only a bolt) within a certain temperature range; salt crystals are only salt crystals in the absence of a solvent, and so on. All this suggests that the independence or priority of parts over wholes must be looked at on a case-by-case basis, and that independence will be a matter of degree.

2.8 Conclusion

In this chapter we have seen that minimal mechanism provides an abstract and widely applicable account of what a mechanism is—one that can provide some theoretical unity to our understanding of the range of things that fall under scientific and commonsense conceptions of mechanism. Let us recap briefly some main conclusions that we can take so far:

1. Mechanisms are particular and compound, made up of parts (entities) whose activities and interactions are located in particular regions of space and time.

2. Mechanisms are always mechanisms for some phenomenon; they produce or underlie the phenomena for which they are responsible. A mechanism's phenomenon will be determined by the organization of its entities, activities, and interactions. Organization runs in two dimensions, the horizontal dimension of spatio-temporal and causal organization, and the vertical dimension whereby the organized activities and interactions of parts constitute the activity of the mechanism as a whole.

3. Mechanisms are related both to systems and to processes. Systems are stable configurations of entities with persistent capacities, so systems are kinds of compound entities. Processes are collections of entities acting and interacting through time. When systems act or interact within their environment, a mechanistic process occurs within the system. Some processes, though, are not activities of systems, but are ephemeral and non-systematic.

4. The parts of a mechanism will typically themselves be systems, whose behaviors and capacities will be determined by the entities and activities of which they are composed. Similarly, the activities and interactions within a mechanism are typically themselves processes involving finer-grained entities and activities. We can express this hierarchical organization of mechanisms by saying that the entities, activities, and interactions within a mechanism are typically themselves mechanism-dependent.

5. The boundaries within mechanisms as well as those between mechanisms and their environment are determinate only with respect to a phenomenon for which the mechanism is responsible. Although the same system may be decomposed in different ways depending upon which behavior of the system has been identified as a phenomenon, these boundaries still reflect objective and mind-independent features of nature.

6. Most mechanisms that scientists study behave in regular but not exceptionless ways; still, one-off or ephemeral mechanisms fall under the scope of the minimal conception of mechanism. Many of the generalizations that receive the honorific title of "law" are just descriptions of the behavior of mechanisms.

7. If the world is in fact made of mechanisms, it has consequences for the philosophical understanding of the ontology of the natural world. New Mechanist ontology is an ontology of compound systems. It suggests that the properties and activities of things must be explained by reference to the activities and organization of their parts. Mechanisms, and with them properties and causes, are particular and local. The New Mechanist view suggests the need for revisions in the way philosophers often think of the concepts of substance, property, and law.

In emphasizing the particularity of mechanisms as things in the world, I have downplayed the importance of generality in the way we understand and represent mechanisms. Even if mechanisms are, from an ontological point of view, particulars, scientists are not typically concerned with single mechanisms, but with classes of mechanisms—mechanism kinds. To understand the role of generality, we must turn from ontology to representation. We must see how mechanisms are modeled.

3

Models, Mechanisms, and How Explanations

3.1 What Are Models?

The only perfect model of the world ... is ... the world itself.   (Paul Teller 2001)

The New Mechanical Philosophy is both a philosophy of nature and a philosophy of science. It tells us something about how the world is, as well as something about how we, particularly through the methods and institutions of science, may come to know that world. In this chapter we shift from questions about mechanisms in the world to questions about how scientists represent those mechanisms.

According to the analysis I will present, the central vehicles for representing the world and its causal structure are models, and in particular mechanistic models. This is not an original idea. Two strands of thinking have emphasized the importance of models in science. The first is the semantic view of theories (Suppe 1977; Winther 2016), which argues that the proper way to understand scientific theories is as families of models. The kinds of models referred to in the semantic view were in the first instance mathematical models of the Tarskian variety, models that stand in formal satisfaction relations to syntactically characterized scientific theories. The second strand is the development of what I will call "the modeling view of science." Whereas the semantic view was primarily in the business of providing an analysis of the nature of scientific theories, and in particular with correcting the idea that theories could be conceived of even ideally as syntactic objects, the modeling view of science sees modeling as a different kind of enterprise, with models functioning as mediators between theories and the world (Cartwright 1983; Giere 1988; 2004; Morgan and Morrison 1999; Teller 2001; Weisberg 2007a; 2007b; 2013; Wimsatt 2007, ch. 6). My own account will largely follow Giere's (1988; 2004; 2006) account.[1]

What is the relationship between modeling and other kinds of scientific activities? This depends upon what you take models to be. There is what we might think of as a narrow use of the term "modeling," where modeling refers to activities involving the construction and investigation of mathematical and computational models. Modeling

---

[1] The relationship between the semantic view of models and scientific modeling is complicated, and I will not attempt to unravel it here. For two interesting attempts, see Teller (2001) and Morrison (2007). I also address the issue in relation to mechanistic models in Glennan (2005).

in this sense is a specialized subdiscipline of many scientific fields. I shall use the term "model" in a broader sense to refer to a whole range of devices scientists use to represent aspects of the world. In a sense, I will concur with the advocates of the semantic view, who say that scientific theories just are collections of models, though I shall have a different account of what models are and of how they are bundled together.

The things that scientists have called models are at first glance an incredibly diverse set—equations, computer simulations, ball and stick models of molecules, scale models of airplanes or organs, animal models, experimental models, and so on. Some of these models are abstract objects while others of them are physical things or systems. Is there anything that these things share that make them models? According to Giere, the crucial feature is that models are always used to represent things. Teller, following Giere, suggests that "in principle, anything can be a model, and that what makes a thing a model is the fact that it is regarded or used as a representation of something by the model users" (2001, 397). There is no intrinsic property that characterizes models; they must be used to represent. Giere (2004, 743) emphasizes the agent-dependence of representation in the following schema:

S uses X to represent W for purposes P.

Here, according to Giere, S can represent an individual scientist, a group of scientists, or the scientific community as a whole, and we can extend that beyond scientists to other cognitive agents. What is key is that representation is something that is done implicitly or explicitly by some individual or collective intentional agent and that the model is always used for some purpose.[2]

Let us call the piece of the world that is modeled (W) the target. How exactly does the agent use the model to represent the target? She does so by means of what Giere (1988) calls a *theoretical hypothesis*. This is a hypothesis that the model resembles the target in some degrees and respects. What degrees and respects are called for depends upon the modeler's purposes.[3] To take a simple example, consider two different models of a Boeing 787 Dreamliner. One is constructed by engineers for the purpose of testing aerodynamic properties of the fuselage and wings in a wind tunnel. Another model is built for the purpose of testing and displaying various interior seating and lighting configurations. These models will be entirely different. The crucial properties for the wind-tunnel model are the shape (but not size) of the wings and fuselage. The crucial properties for the seating model include correct interior dimensions and materials.

---

[2] Weisberg (2013) shows that not all models have actual targets. They may, rather than referring to actual targets, refer to possible or imagined targets, or to no target at all. Weisberg's account, however, starts with targeted models and is in most respects in agreement with Giere. For instance, both insist on the purpose-driven character of model use and the idea that the model-target relation is one of similarity. For the purposes of this chapter, targetless models are not central.

[3] Giere has dropped the term "theoretical hypothesis" from his recent accounts, mostly I believe because he has chosen to focus on the act of representing via the schema above. Weisberg (2013) describes this activity of interpreting the model as "construal." I shall use the term "construal" in this sense, when it is linguistically convenient.

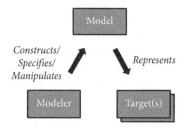

Figure 3.1 The model-target relation.

Figure 3.1 illustrates the relationship between the modeler, the model, and the target(s). The modeler (who can be an individual or collective agent) builds and describes a model through heterogeneous means, including textual descriptions, equations, diagrams, and computer programs. In the case of material models, the construction may be literal, as for instance when a modeler builds a model of a molecule out of balls and sticks or when the engineers construct wind-tunnel models. The models, whether material or non-material,[4] should be distinguished from their descriptions. Different sets of sentences or diagrams may specify or describe the same model. In the case of material models, the model just is whatever material objects and processes in the world make up the model. In the case of non-material models, the model just is the logical/mathematical structures specified by the equations, programs, or diagrams. Non-material models are semantic rather than syntactic entities.

In an illuminating metaphor, Giere (1999; 2004) suggests that models are like maps. Maps will represent a piece of the world, but only to a certain degree of accuracy and in certain respects. There is no one best map of a place, and what counts as a good map will depend completely on the user's purposes. For instance, a map of the London underground represents some basic topological properties of London's transport system—namely the order of stations on the various lines, and the places where lines intersect. It is great for figuring out the various ways that one may transfer on the Tube or the London Overground in order to reach certain stations. It does not tell you anything about the above-ground geographical relationships between stations, like distance and compass orientation. Nor does it tell you about any other features of the terrain besides the stations themselves. Other maps and mapping devices are needed to help us find street addresses, services, and landmarks.[5]

It is notable that many of the things that make for good maps do not have anything directly to do with their representational content, but with evidently pragmatic virtues. Good maps, for instance, are compact and easy to fold. GPS mapping devices are great,

---

[4] I use the word "non-material" rather than the somewhat more common "abstract." I will reserve the term "abstraction" for another concept—that of omitting details. In this sense, both material and non-material models can be more or less abstract.

[5] While models are *like* maps, ordinary maps are literally models in the sense required by Giere; they are objects used by individuals and institutions to represent targets for particular purposes.

so long as you have affordable and reliable network coverage. A person navigating their way around London will choose their tools differently depending upon their project—whether they are walking, driving, or using public transport; whether they are seeking to get to places by the fastest route, the shortest route, or whether they are trying to save money; whether they are looking for museums or friends' houses. The same can be said for scientific models. One's choice of a model organism, for instance, will have much to do with practical considerations like laboratory space, the duration of life cycles, ethical considerations for the treatment of animals, availability and cost of equipment, and so on. Such practical considerations can shape and sometimes distort the representation of target phenomena (Bolker 1995; 2012). Similarly, the choice of kinds of mathematical and computer models will depend crucially upon the analytic tractability of the models, and required computational resources. Good models are models you can use.

3.2 Models and Generality

Figure 3.1 illustrates another important feature of the modeling relationship. Models are our source of generality. While a given model is a particular, a single model can often be used to represent a whole class of targets. The model then is a generalized representation of a class of similar targets, and allows one thereby to make generalizations about those targets.

As an example of this kind of generality, consider two related models—one material and the other mathematical. The squid giant axon is an experimental model that was used by physiologists in the 1930s and 1940s to explore how signals are transmitted along nerve fibers (axons). It was through study of this model system that the British physiologists Hodgkin and Huxley were able to formulate a mathematical model of the action potential, which is the pattern of transmission of depolarization along the length of an axon, the so-called Hodgkin Huxley (HH) model (Hodgkin and Huxley 1939; 1952).[6]

A squid's giant axon is the axon of a very large neuron that is part of the water jet propulsion system in the squid. In comparison with most nerve cells, the giant axon is very large, with a diameter of up to one millimeter and lengths of several centimeters. Because of its large size, the giant axon is much easier to manipulate experimentally than axons from other neurons, and the empirical study of these axons permitted scientists to understand the general mechanisms by which nerves transmit electrical signals along their axons. Using laboratory equipment available in the 1930s and 1940s, Hodgkin, Huxley, and others were able to remove axons from squid and to test their behavior in response to changes in the electric potential inside and outside the

---

[6] This model has been much discussed in the philosophy of neuroscience literature (Craver 2007; 2008; Bogen 2008a; Weber 2008), with particular attention to the question of the extent to which the HH model is explanatory, and whether it constitutes a nomological or a mechanistic explanation.

membrane. The size of the giant axon was large enough to permit the insertion of electrodes inside the axon. These electrodes were parts of "voltage clamps" that permitted the measurement of ion currents in the axon under different voltage conditions.

The giant axon is a good example of a material and experimental model. Physiologists perform experiments on individual squid giant axons, carefully prepared. They do so, not because of an interest in squid giant axons per se, but because these axons are taken to be representatives of neurons generally. Of course in certain respects the giant axon is decidedly not like other neurons, but it is its most distinctive and unusual feature— its giantness—that makes it such a valuable experimental model in the first place. Additionally, the squid giant axon is not studied in its normal physiological context, but rather in a highly artificial laboratory context in which it has been removed from the squid, attached to electrodes, put in a solution, and subjected to differing voltages. But the hypothesis (and this is a theoretical hypothesis) is that these experimental setups do not alter the basic behavior of the squid axon, but simply permit manipulation of the axon in ways that will elucidate that behavior. Moreover, it is assumed that the squid giant axon is a good model of the action potential in neurons generally. That is to say, it is another theoretical hypothesis that with respect to the basic dynamics of the action potential, squid giant axons are relevantly similar to other axons in squid and to axons in other animals.

Hodgkin and Huxley used the results of experiments with the squid giant axon to develop their mathematical model. The HH model describes the action potential by means of the following differential equation:[7]

$$I = C_M dV/dt + G_K n^4 (V - V_K) + G_{Na} m^3 h (V - V_{Na}) + G_l (V - V_l)$$

The model describes the triggering conditions for depolarization as well as the general shape of the curve representing changing potential difference over time. The four terms of the equation have a physical interpretation: The first represents the capacitance of the membrane, the second the potassium current, the third the sodium current, and the fourth a "leakage current" representing other ion currents.

While the model was originally presented as a model of the action potential in this particular kind of neuron, its import depends upon the plausible hypothesis that the equation (with suitable choice of parameters) describes the electrochemical behavior of neurons generally. Writing about the range of applicability, Hodgkin and Huxley remark:

The range of phenomena to which our equations are relevant is limited in two respects: in the first place, they cover only the short-term responses of the membrane, and in the second, they apply in their present form only to the isolated squid giant axon.

(Hodgkin and Huxley 1952, 541)

---

[7] This form of the equation is taken from Craver (2006, 363).

But they go on to observe that with suitable changes of parameters, the equations may describe the behavior of many tissues:

The similarity of the effects of changing the concentrations of sodium and potassium on the resting and action potentials of many excitable tissues…suggests that the basic mechanism of conduction may be the same as implied by our equations, but the great differences in the shape of action potentials show that even if equations of the same form as ours are applicable in other cases, some at least of the parameters must have very different values.   (542)

The generality of both the squid giant axon model and the HH model derives from the hypothesis that the model resembles more than one target. Strictly speaking, though, we should note that the squid giant axon model is not a single experimental model, as many different particular squid axons have been used to collect experimental data. In the case of the HH model, we do have a single mathematical model that can represent countless nerve cells. The model may be made more specific by specifying values of parameters, but in all cases the mathematical model is not meant to stand for one neuron, but for entire classes of neurons.

Not all models are general in this sense. Models of one-off systems or processes are models that stand in one-to-one relations to their targets. The map of the London underground is a model of the London underground and not undergrounds generally. Similarly, the sorts of models that people construct to understand historical events can be about just a single event. For instance, when one constructs a model (physical, verbal, mathematical) to account for, say, the crash of a plane or train, that model is supposed to represent the processes leading up to that very particular event. Historical narratives, as we shall discuss later in this chapter, are models of this kind.

3.3 Models and Theories

If Giere and other advocates of the model-based view of scientific representation are right, what role is left for theory? A helpful way to think of theories is as toolkits for building models (Cartwright 1995; Teller 2001). Giere (2004; 2006) has advocated a similar view, under the name of "principles." To begin with a simple and standard example, Newtonian mechanics and the theory of gravitation provide a set of principles for constructing models of specific systems. For instance, for a model of the solar system, the theory tells you to start with initial locations, velocities, and masses for the various celestial bodies. The theory then tells you how to calculate the instantaneous force for each object within the system, which is the sum of the gravitational forces generated by each pairwise relation with other objects in the system, and from this you get instantaneous rates of accelerations for each object in the system. One can build similar models of other planetary systems, of moons, of satellites, etc. Gravitational models can be extended by the addition of further principles governing other Newtonian forces. For instance, a model of the motion of a terrestrial projectile like a cannon ball might add an additional force representing air resistance. This would require some sort of theory of air resistance.

Theories on this view are often collections of generalizations or laws, but the laws are not understood as being literally true descriptions of anything, but rather principles that can be applied, along with approximations and idealizations to particular systems, or to classes of systems. The question of which principles to apply in developing the model is influenced by the interests and purposes of the modeler. For instance, whether air resistance is a factor that needs to be modeled in an account of projectile motion will depend both on how accurate the model must be for whatever purpose the model is being deployed as well as on how significant a force air resistance is, which will depend upon the particulars of the objects being modeled.

The case of Newtonian mechanics is in many respects unusual because of the clarity of the rules for applying principles in the construction of models. The additivity of Newtonian forces makes it easy to model the combined effects of many forces. In lots of modeling activities, however, the various principles appealed to will be heterogeneous, and there will be no fixed rules for how to model their interactions. As an example, consider climate modeling. Lloyd describes the relation between climate models and theories as follows:

> There is no general theory of climate that takes all the complicating factors affecting climate into account and calculates what the effects on climate change will be on global temperature change, on precipitation, on wind, on pressure change, or on any other significant climate variable. So the climate scientists combine general pieces of theory from fluid dynamics, thermodynamics, and theories regarding radiation with detailed models of how the equations are applied to individual parts of the climate system. Details about the ice, vegetation, soil, and water vapor and the ornate interconnectedness of systems are represented in these global climate models.   (Lloyd 2010, 972)

Here we notice that climate models are not based on any single theory, but use various theories as tools to model a particular system. Those theories can be thought of as describing various kinds of entities and activities involved in a heterogeneous and multi-level mechanism.[8]

We can relate this account of theory or principles back to our account of mechanisms in the following way. Theories provide generalized accounts of kinds of activities, and the sort of entities that can and do engage in them. So, for instance, gravitation or convection are kinds of activities. Gravity requires massive bodies; convection requires fluids. A model, or at least a mechanistic model, as we shall define it below, takes these principles and applies them to a particular case or class of similar cases. This notion of theory does not give theories priority over the models that use them.

---

[8] For some case studies showing how heterogeneous theories are integrated into multi-level mechanistic models see, e.g., Craver's (2007, ch. 7) account of the mechanism of long-term potentiation and Darden and Craver's (2002) discussion of mechanisms of protein synthesis.

The entities and activities referred to in these general theories or principles will typically be mechanism-dependent.

3.4 Mechanistic Models and How Explanations

The account of modeling I have described above is meant to apply to models generally. Now I shall turn to a particular variety of model I shall call a mechanistic model. Models are representations, and mechanistic models are representations of mechanisms. They are not just any representations, but representations that describe (in some degrees and respects) the mechanism that is responsible for some phenomenon. As such, mechanistic models are to be contrasted with phenomenal models, which may be useful for describing and predicting the behavior of a mechanism, but do not explain how that behavior is produced.[9]

Mechanisms are a collection of entities whose activities and interactions are so organized as to be responsible for a phenomenon. Accordingly, we may think of a mechanistic model as having two parts—a model of the phenomenon, and a model of the mechanism that is responsible for that phenomenon. We can call these two parts of the model the phenomenal description and the mechanism description (Glennan 2005). While the phenomenal description is a phenomenal model that is useful for describing and predicting the mechanism's behavior, it is the mechanistic model, which characterizes the entities or parts and their activities and operations, that actually explains (Glennan 2002a; Bechtel and Abrahamsen 2005; Craver 2006). While we may be able to distinguish conceptually between these two elements of a mechanistic model, in practice the phenomenal and mechanistic elements are often not separated out; also, models will invariably be idealized and partial, representing the various aspects of the mechanism only in certain degrees and respects.

Many advocates of mechanistic explanation (Bogen 2005; Craver 2006; 2007; Kaplan and Craver 2011; Kaplan 2011) have emphasized the distinction between phenomenal and mechanistic models in order to separate the descriptive and explanatory function of models, but sometimes separating these functions out is less easy than it might appear. Let us consider some examples of phenomenal models to see why. A classic example of a phenomenal model is a Ptolemaic model of celestial motions. Ptolemaic models can be very good at summarizing the apparent motion of the planets through the sky, even though the basic assumptions of the model, like the assumptions that the earth is at the center of the solar system and that planets travel on paths compounded out of multiple circular orbits, are not true. Notice, though, that there is nothing intrinsic to the Ptolemaic model that requires one to interpret it as a phenomenal model rather than a mechanistic one. It is quite possible to interpret

---

[9] My focus here is on the contrast between phenomenal and mechanistic models, but I do not claim that these are the only sorts of models. Weisberg (2013) provides examples of a wide array of models and strives for a classification that is complete—though I will not judge how successful he is in this attempt.

the epicycles and equants as representing real parts of a celestial mechanism, and in times past geocentric astronomers interpreted at least some parts of Ptolemaic theory as characterizing a mechanism (Kuhn 1957).

Contrast this to the much-discussed case of the HH model of the action potential. Bogen (2005) and Craver (2006) emphasize the phenomenal aspects of the HH model. It describes how potassium and sodium currents change, but does not explain how it works. In interpreting it this way, they cite Hodgkin and Huxley themselves, who explicitly characterize their results as descriptive. Hodgkin and Huxley emphasize that their data does not support a claim that their "equations are anything more than an empirical description of the time-course of the changes in permeability to sodium and potassium" (1952, 541).

Notwithstanding Hodgkin and Huxley's modesty, the HH model is not a purely descriptive and phenomenal model. It is rather a partial and incomplete model of the mechanism. As Craver puts it:

> They began their project with a background sketch of a mechanism. They knew some of its entities and activities. They knew that action potentials are produced by the movement of ions across a lipid membrane. They knew that action potentials are produced by depolarizing the cell body ... And they knew that the shape of the action potential ... could possibly be produced by the voltage dependent activation and inactivation of the membrane conductance for specific ions.   (2007, 115)

The point here is that the terms of the HH model are physically interpretable as representing some of the entities, activities, and organization of the mechanism; if the model is so interpreted (via theoretical hypotheses), and if those hypotheses are correct, then to this extent the model explains. It does not of course explain everything, and in particular it provides no account of the mechanism responsible for changes in resistance across the membrane, but the fact that the explanation is partial does not make the model purely phenomenal.[10]

The distinction between a phenomenal and a mechanistic model is the distinction between a model that describes the phenomenon and one that explains it. A phenomenal model characterizes what the phenomenon is—often no mean achievement—while the mechanistic model shows how it comes about. Two points about this distinction bear repeating. The first is that no model can be judged to be phenomenal or mechanistic absent the modeler's interpretation. Models are only models when they are used to represent something, and it is the modeler's act of interpreting the model— asserting the similarity between some aspect of the model and some aspect of the

---

[10]   Craver has two largely parallel discussions of the HH models (2007, ch. 4; 2006). In these papers there seems to be some vacillation over the question of whether the HH model explains. The question depends upon whether we characterize the HH model as a phenomenal model or as a mechanism sketch. In 2006 (p. 356) he observes that Hodgkin and Huxley considered their model to be phenomenal, and he calls it a "phenomenological sketch." But elsewhere (especially in 2007) he refers to the HH model as a mechanism sketch. The point here is that sketchy mechanistic models are different from phenomenal models, and sketchy models explain, albeit sketchily.

target—that determines what kind of model it is. The second point is that models, so interpreted, may be more or less mechanistic. Some terms, for instance, in a mathematical model may be taken to refer to entities and activities that are responsible for the phenomena, whereas others may just be curve-fitting parameters.

Mechanistic models are the vehicles for mechanistic explanation. A mechanistic model characterizes both the phenomenon to be explained and how the organized activities and interactions of some set of entities produce or underlie that phenomenon. So mechanistic models show how the phenomenon is caused and constituted by a mechanism. It is for this reason that I call mechanistic explanations *how explanations*.

While mechanistic models explain, I should emphasize that offering up a mechanistic model is not the only way to explain a phenomenon. There is an ancient tradition that understands explanations to be answers to why questions, and one very important way to explain why some phenomenon occurs is to show how that phenomenon occurs. It is not the only way. Sometimes you can explain why something is without showing how it happened. I shall discuss such cases and explore the connection between mechanistic explanations and other forms of explanation in the last chapter of this book, but for this chapter the focus will be on mechanistic models and how explanations.

To echo Teller's remark that serves as the epigraph for this chapter, the only perfect model of a mechanism is the mechanism itself. This means that no model will tell you exactly how a mechanism works. In the remainder of this chapter, I shall turn to the abstraction and idealization required for mechanistic models, and the different senses of "how" involved in mechanistic explanation.

3.5 How-Possibly, How-Actually, and How-Roughly Models

One natural way to read the literature on mechanistic explanation is that mechanistic explanation eschews abstraction. What distinguishes mechanistic explanations from non-mechanistic descriptions or accounts is that mechanistic explanations fill in the details. They tell you what the parts are and what they do. So, for instance, Bechtel and Richardson (1993) suggest discovering mechanisms is a matter of decomposition and localization of functions. Similarly, Craver sometimes describes scientific progress as a matter of filling in the mechanistic details behind filler-terms in mechanism sketches (2007, 113–14). Again, Kaplan and Craver (2011) urge that dynamical models in neuroscience only explain insofar as they are interpretable as saying something about the entities and activities that make up mechanisms.[11]

---

[11] Levy and Bechtel (2013) have suggested that there is a systematic difference between the New Mechanists, with Craver and Darden in particular always arguing for "the more concrete the better" approach to explanation (see also Chirimuuta 2014). Craver (personal communication) believes his and Darden's opposition to abstraction is overstated, and that he has made numerous remarks in print on the necessity and value of abstraction and idealization. See, for instance, Craver (2013, 49–51) and Craver and Darden (2013, 32–4).

Craver's concerns with excess abstraction are epistemic. In his view, a crucial aspect of mechanistic explanation is that it constrains more abstract functional explanations (Craver 2007; Piccinini and Craver 2011). Functional explanations contain lots of black boxes and filler terms, and by trying to locate the black boxes and figure out what goes in the filler terms, one gets a much better idea of whether the putative functional explanation is correct. Craver makes two "normative distinctions" regarding potentially explanatory models. The one concerns epistemic plausibility: models can be arrayed on a continuum from how-possibly to how-plausibly to how-actually. The other concerns "sketchiness"—on a continuum from mechanism sketches to mechanism schema to "complete mechanistic models." These distinctions are normative because "progress in building mechanistic explanations involves movement along both the possibly-plausibly-actually axis and along the sketch-schema-mechanism axis" (Craver 2007, 114).

These axes are not, however, as tightly linked as Craver sometimes seems to suggest (Gervais and Weber 2013). While one cannot doubt that moving from an explanation of how possibly to one of how actually constitutes scientific progress, it is not always the case that moving to less "sketchy" models does. If by sketchiness one means abstraction, for some purposes a sketchier model may be a better one. This may be both because the more abstract model can apply to more targets, and also because the more abstract model may capture the central explanatorily relevant features in a way that is independent of the gory details (Kitcher 1984).

There are two ways to understand what makes a model a how-possibly model.[12] On the one hand, the model could be an (epistemically) possible model of the actual mechanism. The model in this instance is construed as a hypothesis or conjecture about the mechanism responsible for some phenomenon. Should the conjecture be true, then the model would actually explain the phenomenon, but we are uncertain whether the conjecture is true. If this is the case, there is nothing that intrinsically distinguishes a how-possibly from a how-actually model. The difference rests solely on whether or not the theoretical hypotheses concerning the model-target relation are known to be correct. To call a model a how-possibly model in this sense is to say something about its warrant. Craver's interpretation of how-possibly models seems to follow these lines:

How-possibly models (unlike merely phenomenal models) are purported to explain, but they are only loosely constrained conjectures about the mechanism that produces the explanandum phenomenon. They describe how a set of parts and activities might be organized such that they

---

[12] The term 'how-possibly explanation' originates with Dray (1957) in his discussion of historical explanation. Dray was responding to a rather different set of questions than Craver, and was chiefly concerned with showing a sense in which one could explain in history without reference to covering laws. More recently, there have been a number of discussions of how-possibly explanations in biology (Resnik 1991; Forber 2010; Reydon 2012), though to my knowledge Craver is the first to develop a notion of how-possibly *models*.

exhibit the explanandum phenomenon. One can have no idea if the conjectured parts exist and, if they do, whether they can engage in the activities attributed to them in the model.

(Craver 2006, 361)

Notice how Craver describes the model as a "conjecture" and how one may have no idea whether the conjectured parts or activities exist. How-possibly models on this account are just possible how-actually models, with minimal subjective warrant.[13]

A second way of thinking about how-possibly models is to say that they are not conjectures about how a mechanism actually works, but they are rather models that represent something other than the actual. They are models about "mechanistic possibilities." Mechanistic possibilities are in a certain way analogous to physical possibilities. On many accounts, to say that something is physically possible is to say that it is consistent with physical law. For instance, it is physically possible for an object to travel close to the speed of light, but it is not possible for it to go faster than the speed of light. Claims about physical possibility thus do not explain actual events in the world, but rather the kinds of events and states of affairs that can occur in the actual world. Analogously, claims about mechanistic possibility do not refer to actual events or states of affairs in the world, but rather claim that a certain kind of mechanism can or cannot be responsible for some kind of phenomenon. Michael Weisberg describes some instances of this sort of strategy in biological modeling:

Some modelers construct models simply to explore or illuminate a hypothesis. For example, population biologists often examine very simple models of sexual and asexual reproduction in order to better understand the evolution of sex...These models are not intended to describe any actual organism; they are far too simple for that. Their importance lies in helping us to understand very general facts about the differences between sexual and asexual reproductive systems....

It is also possible to study a model of a phenomenon that is known not to exist. A. S. Eddington once wrote: "We need scarcely add that the contemplation in natural science of a wider domain than the actual leads to a far better understanding of the actual"...Agreeing with Eddington, R. A. Fisher explained that the only way to understand why there are always two sexes involved in sexual reproduction is to construct a model of a three-sexed sexually reproducing population of organisms...   (Weisberg 2007b, 223)

These models are not meant as conjectures about mechanisms in actual populations, but instead are studied in their own right to understand mechanistic possibilities.

It will clarify matters to place the distinction between how-possibly explanations and how-actually explanations in the context of Giere's formulation of the model-target relation. The model-target relation we have seen is determined by a construal of the model in the form of a theoretical hypothesis. To say of a model that it is a how-possibly model in Craver's subjective sense is to say something about the epistemic

---

[13] Craver might argue that there is some intrinsic difference, because how-possibly models tend to be sketchier and less rich than those approaching the how-actually end of the spectrum. But this is at best a tendency, and it is quite possible to have a model rich in mechanistic detail that has got the detail wrong. See Gervais and Weber (2013) for discussion.

status of this hypothesis. How-possibly models are not well confirmed. In this sense, Craver is exactly right about the norm that we should move from how-possibly toward how-actually models; but this just means we should prefer better confirmed models. The epistemic status of the hypothesis is independent of its representational content, so the very same model we classify as how-possibly may turn out to be a model of how the mechanism actually works.

But, as Weisberg's examples show, how-possibly models are not always used as hypotheses about how a mechanism actually works. Sometimes the point is to show that some hypothesized mechanism could not produce the phenomenon in question. Take, for instance, blending inheritance, the mechanism Darwin hypothesized to underlie transmission of traits between generations. The idea of blending inheritance was that the causal basis of traits (like height of a plant or color of a flower) would be transmitted during reproduction, and that the resulting offspring would tend to have traits that were intermediate between the parents. The problem with this mechanism, which simple models can demonstrate, is that it will tend to destroy the variation in traits within a population that is required for natural selection to act upon. Such models suggest that natural selection plus blending is an inadequate account of evolutionary mechanisms, as Darwin's critics realized. In so doing, these models suggest a constraint on possible mechanisms for inheritance—a constraint met by the actual mechanism discovered and elaborated by Mendel and his successors in genetics.[14]

But, returning to models with actual targets, we must explore further the relationship between how-possibly mechanisms and their potential targets. As I have put it above, a how-possibly model in this sense is just a possible model of an actual mechanism. But this way of putting it oversimplifies the model-target relation. If Giere's account is correct, the relationship between model and target is always one of similarity in degrees and respects. If this is so, we cannot divide models into those that do hold of the target and those that don't, i.e., we cannot say model A represents a possibility that is not actual, while model B represents a possibility that is actual. Models represent in degrees and respects, or more or less. If we hold a model to a less strict similarity requirement, it may succeed in representing an actual target, if only roughly. When a model (actually) does this, we can call it a *how-roughly* model.

Consider as an example a simple Copernican model of the solar system. The big idea of Copernican astronomy (the Copernican revolution if you will) is that the earth is a planet, and in particular the third planet from the sun. The simplest Copernican model is a model that supposes that the solar system consists of a number of planets traveling on circular orbits at a constant speed (different for each planet) around the central sun. This model resembles the actual solar system in certain

---

[14] As Wimsatt (2002) shows, the story about blending inheritance is not as simple as this. The decline of variation in a blending model could be counteracted if selection acted against blending, or if mating was non-random. Wimsatt also notices another important heuristic purpose of how-possibly models. They can help us identify potential mechanisms in domains different from those for which they were originally formed.

degrees and respects. In particular, it saves some of the major qualitative phenomena like the periodic retrograde motions of the planets, and it explains why the interior planets of Venus and Mercury are never seen far from the sun. It also correctly characterizes some important features of the mechanism that produces the apparent motions of the planets—in particular, the central position of the sun and the order of the planets. Of course, in more quantitatively precise terms the simple Copernican model is a failure—a failure because the uniform circular orbits are at variance with the true (or almost true) elliptical orbits characterized by Kepler and Newton.

There are two different ways we might construe this simple model. If we construe it loosely, by saying that it is qualitatively similar to the actual solar system with respect to the location of the sun, planet order, and direction of orbits, then we have a good how-roughly model of the actual solar system. If, on the other hand, we construe it more tightly as explaining more exact apparent locations of planets over many years, then the theoretical hypothesis is simply false; the model does not reproduce the phenomena at the required level of similarity.

As another example of how-possibly and how-roughly models, consider van Gelder's well-known argument for a dynamical systems approach to cognition (van Gelder 1995). His title "What might cognition be, if not computation" suggests immediately that he is offering a how-possibly model. He wishes to argue that dynamical non-computational approaches to modeling cognitive processes could possibly explain many cognitive phenomena. To do so, van Gelder draws a distinction between two approaches to understanding cognitive phenomena. On the one hand, a computational approach models cognitive processes as sequential processes following rules that manipulate internal representations. The dynamical systems approach van Gelder would have us consider is one in which there are not discrete representations or rules, but where there are continuous interactions between parts of the system whose dynamics can typically be described by non-linear differential equations.

Van Gelder illustrates the contrast between dynamical and computational approaches by comparing two approaches to modeling human decision-making. He contrasts the dynamical approach with the classical approach of von Neumann and Morgenstern, an approach that treats decision-making as a process of calculating expected utilities. According to van Gelder, this approach is an instance of a computational approach, and in all of its variants it suffers from a number of flaws, which he summarizes as follows:

(1) They do not incorporate any account of the underlying motivations that give rise to the utility that an object or outcome holds at a given time.

(2) They conceive of the utilities themselves as static values, and can offer no good account of how and why they might change over time, and why preferences are often inconsistent and inconstant.

(3) They offer no serious account of the deliberation process, with its attendant vacillations, inconsistencies, and distress; and they have nothing to say about the relationships that have been uncovered between time spent deliberating and the choices eventually made.

(van Gelder 1995, 359–60)

Van Gelder's objections to the von Neumann and Morgenstern approach point to both phenomenological and mechanistic deficiencies. Phenomenologically, the models fail to appropriately characterize certain patterns in decision-making behavior—its temporal character and its oscillations. Mechanistically, they "do not incorporate any account of the underlying motivations that give rise to [utilities]" or "how and why they might change."

In place of classical computational models, van Gelder recommends we consider an dynamical systems alternative, such as motivational oscillatory theory or MOT (Townsend 1992). Simple dynamical models construed using MOT exhibit at least qualitatively some aspects of decision-making phenomena that elude classical models. They show, for instance, how motivations can be modified and how there can be a temporally extended set of oscillations before the system comes to a decision. These models are how-roughly models, which, in comparison to the cognitive phenomena they represent, are quite simple and idealized. They are like the simple Copernican model of aspects of planetary motion. But because they get right certain features of the phenomena that have eluded computational models, van Gelder claims there is good reason to believe that in some respects a MOT model is a good model of human decision-making mechanisms. Such models make no claim to be realistic neurologically, and do not replicate decision-making patterns of actual human agents. Nonetheless, they do replicate certain features of these processes better than classical computational models, and this provides some evidence for the view that these models roughly represent certain aspects of mechanisms that actually are responsible for human decision-making.

Notice, though, that the mere fact that a model like MOT describes aspects of a mechanism's phenomenon does not guarantee that the model explains those phenomena, even roughly. To explain, features of a model like MOT must map, to some degree and in some respects, to the organized entities and activities that are actually responsible for the kind of phenomena MOT represents (Kaplan and Craver 2011; Kaplan and Bechtel 2011; Kaplan 2017; Zednik 2011).

3.6 Abstraction and Idealization in Mechanistic Models

The degrees and respects formulation of the model-target relation entails that any model will be to some degree abstract and idealized. Abstraction is the omission of detail from a representation or description of a target, while idealization introduces features into a representation or description that distort features of the target. Abstract models are true so far as they go, while idealized models are in some ways false.[15] This is certainly true for how-roughly models, but it is true also for models that are far along the spectrum toward "complete" models of a phenomenon. In this section, we shall

---

[15] For probing discussions of the nature of abstraction and idealization see McMullin (1985); Batterman (2001); Weisberg (2007a).

explore some examples of abstraction and idealizations in mechanistic models to get a better handle on how they work.

Abstraction in mechanistic models derives chiefly from abstraction in the characterization of the entities, activities, and interactions that are responsible for the mechanism's phenomenon. Consider the mechanistic process by which I got myself to work. Most abstractly, I traveled in a vehicle from point A to point B. Less abstractly, I might describe the entity involved as a bicycle, or as a touring bicycle, or as my green Surly Long Haul Trucker, etc. And less abstractly, I can describe the activity as cycling, and indeed as a particular sort of cycling that one can do on paved trails with bikes equipped with derailleurs and caliper brakes, and so on. I can also break the trip into segments, describe component activities like pedaling and braking, and so on. All of these descriptions, from the most abstract to the most concrete, are true. It is just the more concrete descriptions tell one more and more about the specific entities and activities involved.

Contrast this with idealization. In an idealization, the representation deviates from the target so that some features of the target are different from the representation. The representation is in some respects false. Consider again my commute. If I want to give a detailed account of the relationship between my speed and my pedaling along the ride, I would likely describe the trip as consisting of a series of segments, different ones for each time I changed gears. Such a representation would likely treat each segment as contiguous with the next and treat gear changes as instantaneous; but gear changes are not instantaneous, so a description that treats them as such is idealized.

We will often speak of models as being abstract and/or idealized, and mostly this is harmless, but we should observe that strictly speaking, the idealization does not lie in the model itself, but in the construal by which the model is used to represent the target. For instance, in Figure 3.2, we have a diagram that is a model of a bike's drivetrain. It only becomes abstract when we construe it as a model of a particular bike. When we do, we will notice that much is left out (like the control cables, the wheel, etc.) and some things (like the number of cogs in the freewheel) may simply be incorrect.

As we have seen, Craver has characterized progress in mechanistic explanation as a process of moving from mechanism sketches to complete mechanistic models (see also Machamer, Darden, and Craver 2000). While Craver makes clear at points that there are few if any complete mechanistic models, he maintains that such models are an explanatory ideal. This is a version of what Teller has called the ideal-model model. My question about the ideal-model model is this: is this model itself a good idealization?

The answer depends I think upon the kind of mechanism. There are some mechanisms for which we can arrive at reasonable approximations of complete models, while there are others for which any model we actually build or describe will not remotely approximate the ideally complete mechanistic model. Given that I believe that even in these latter cases we do manage to build models that are explanatory, it will be important not to base our account of scientific explanation on the ideal-model model.

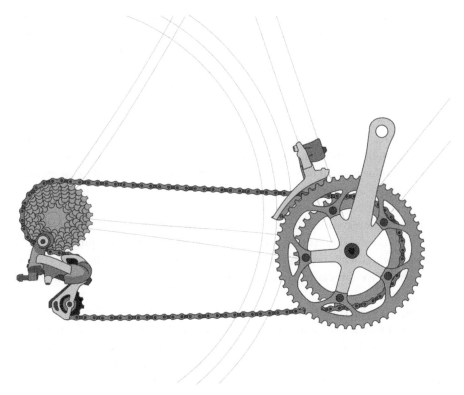

Figure 3.2 Drivetrain for a derailleur bicycle. By Keithonearth (Own work) [CC BY-SA 3.0 (http://creativecommons.org/licenses/by-sa/3.0) or GFDL (http://www.gnu.org/copyleft/fdl.html)].

What might a complete mechanistic model look like? According to Craver, an ideally complete model would "include all of the entities, properties, activities, and organizational features that are relevant to every aspect of the phenomenon to be explained" (Craver 2006, 360). There is an important first step implied by Craver's characterization, which is that even a complete model is a complete model only of some phenomenon that the mechanism produces. Thus even a complete model abstracts from the concrete entities and activities in the world to isolate some phenomenon of which the scientist seeks to explain.

As an example of something that does reasonably approximate a complete model, let us consider a model partially represented in Figure 3.2 that represents the drivetrain of a bicycle using front and rear derailleurs. This diagram pictures most of the essential parts of the drivetrain, but we would need to add to it the cables and shifters that control the derailleurs as well as the hub and wheel to which the freewheel is attached. We would need to fill out the model by providing various details (likely in propositional form) concerning the number of teeth in the gears, the dimensions of various components, and so on.

The drive system on my bicycle actually does many things, but to build a model one must start by isolating some aspects of what the bike is doing and characterize them as the phenomenon to be explained. Let us focus on the following phenomenon: the rate of rotation of the rear wheel (w) is a function of three parameters: the rate of rotation of the crank (c), the setting of the front shifter (f), and the setting of the rear shifter (r): $w = \phi(c,f,r)$.

We are in a position to give a pretty complete model of this phenomenon: The crank set drives the freewheel via the chain. The chain can be moved from sprocket to sprocket both on the crankset and on the freewheel by means of the derailleurs, and the derailleurs are in turn controlled by cables that are attached to the shifters. The wheel's rate of rotation is determined by the ratio between the number of teeth on the front and rear sprockets. On my bike the crank has three sprockets with 26, 36, and 48 teeth. The rear cassette has nine sprockets with 34, 30, 26, 23, 20, 17, 15, 13, and 11 teeth. So, on my bike, in the lowest gear (1-1) one turn of the crank produces 26/34 turns of the wheel, while in the highest (3-9), one turn of the crank produces 48/11 turns of the wheel. We can easily use these gear ratios to characterize the phenomenal description ϕ.

A complete model of these basic kinematic features of bike drivetrains is well within the reach of bicycle mechanics. We can identify all of the parts, describe the inter-actions between them, and calculate gear ratios. We can also generalize the model to a variety of bikes with different cranksets and cassettes by parameterizing on the gear ratio. One thing to emphasize about this "complete" model is that the model is only complete at a single level of the mechanism. It explains how the wheel moves in terms of the interactions between parts of the drivetrain. It does not, however, explain why those parts behave as they do. For instance, chains interact with sprockets as the teeth on the front sprocket drag the chain and the teeth on the rear sprocket are dragged by the chain. To explain these interactions, one must refer to the properties and capacities of the parts. In this example, the chain must be flexible and the links must be of a size that will allow them to grip and release the teeth as they travel over the surface of the sprocket and the rear derailleur must be spring-loaded in a way that allows it to main-tain a chain that is taut enough to keep the chain attached to the sprockets. There is a mechanical explanation for how these components work, but I do not count it against the completeness of the drivetrain model that it does not include explanations of the behavior of the parts. That is another phenomenon, for which we would require another mechanism and another explanation. (If we were to require of a complete mechanistic model that it include not just all entities and activities, but all entities and activities of the sub-mechanisms, then our model would ultimately be given in terms of microphysics, and we would never have anything remotely approximating a complete model.)

For simple behaviors of certain machines like the rate of turn of the bike's drivetrain, it does seem possible to give an approximation of an ideal model. My concern, however, is that this is in fact an atypical case. Mechanisms often have too many constituents and their behaviors are too dynamically complex for us to even reasonably approximate a complete explanation. Such complexity arises even in relatively simple mechanisms like the bike. Consider some more complex phenomena for which the bike's drivetrain is responsible. The first phenomenon we discussed was purely kinematic: what is the rate of rotation of the rear wheel with respect to the crank for a given setting of the gear levers? Someone truly interested in the bike's performance would be interested in some much more complicated questions about the drivetrain. In particular, they would be interested in things like the energy efficiency of the drivetrain. How much force does it take to turn the crank under various conditions? How efficient is the transmission of energy from the crank to the rear hub? How much energy is lost during shifting? These are quite complicated physical questions—presumably the sorts of questions that concern engineers designing high-performance bikes. To answer such questions would minimally require much greater attention to properties of the parts, and it would require us to identify further parts that did not make a difference to the basic kinematics of the gear ratios. Suddenly bearings and lubricants come into play, as does the distance from the chain to the hub in various gear ratios, and so on.

While an engineer, seeking to understand the dynamical characteristics of a high-performance bike, will already fall very short of being able to give an even approximately complete mechanistic model of a bike, those complexities pale by comparison to those found in many of the mechanisms of interest to natural and social scientists. To illustrate the challenges of idealization and abstraction in some of these cases, let me turn to a few examples of mechanistic models in cognitive science and neuroscience.

Paul Thagard has been one of the most vocal advocates of a mechanistic approach to the study of human cognition. He endorses the idea that mechanisms are multi-level, and that understanding mechanisms is a matter of identifying entities, organization, and interactions. He believes that the phenomena studied by the cognitive sciences require the integration of mechanistic models at four levels, the molecular, the neural, the cognitive, and the social (Thagard and Kroon 2006). Thagard's philosophical views have been informed not only by his review of developments in cognitive science, but by work he has done in collaboration with psychologists and computer scientists on the construction and testing of mechanistic models of aspects of human cognition, with a focus on the role of emotions in human cognition.

Much of Thagard's and his colleagues' work has utilized variations of a highly idealized model of affective cognition called HOTCO (for "Hot coherence"). This model employs simple connectionist networks to model human decision-making, using what Thagard calls an emotional coherence approach. The basic idea of coherence models is that humans make decisions in such a way as to maximize the coherence of their (often competing) beliefs, desires, values, and other cognitive states. These cognitive states are represented by nodes in a connectionist network, with the strengths of the links between the nodes (positive or negative) indicating the degree to which one node coheres or fails to cohere with the next. The HOTCO model supplements the basic coherence model by adding emotional valences. Thagard summarizes the relation between coherence and emotional coherence in this way:

In the theory of coherence…elements have the epistemic status of being accepted or rejected. We can also speak of degree of acceptability, which in artificial neural network models of coherence is interpreted as the degree of activation of the unit that represents the element. I propose that elements in coherence systems have, in addition to acceptability, an emotional valence, which can be positive or negative. Depending on the nature of what the element represents, the valence of an element can indicate likability, desirability, or other positive or negative attitudes. For example, the valence of Mother Theresa for most people is highly positive, while the valence of Adolf Hitler is highly negative.   (Thagard and Kroon 2006, 19)

On the emotional coherence theory Thagard proposes, there are two features of an element—how much you accept them and how much you like them, and decisions are made in a way that depends both upon minimizing incoherence and maximizing positive valence. Thagard describes the basics of the HOTCO model as follows:

units have valences as well as activations, and units can have input valences to represent their intrinsic valences. Moreover, valences can spread through the system in a way very similar to the spread of activation, except that valence spread depends in part on activation spread. An emotional decision emerges from the spread of activation and valences through the system because nodes representing some actions receive positive valence while nodes representing other actions receive negative valence. The gut feeling that comes to consciousness is the end result of a complex process of cognitive and emotional constraint satisfaction. Emotional reactions such as happiness, anger, and fear are much more complex than positive and negative valences, so HOTCO is by no means a general model of emotional cognition. But it does capture the general production by emotional inference of positive and negative attitudes toward objects, situations, and choices.   (20)

Thagard's general description is already sufficient to show us that this model is very far from a description of an actual cognitive mechanism. The units that are the nodes in the model's network are high-level cognitive states like beliefs, goals, or decisions that cannot be related in any clear way to neurological entities. The algorithms describing the relationship between nodes use simple mathematical formulas that do not correspond to known neurological activities. As Thagard notices, emotional connections between high-level states are represented by single numerical values, and there is no distinction in the model between kinds of emotions.

The HOTCO model is very far from identifying actual entities, activities, and organization in the brain. It is, for instance, far sketchier that the very abstract model represented by Watson and Crick's "Central Dogma of molecular biology" (see Machamer, Darden, and Craver 2000 for discussion). That sketch, while idealized and in many cases highly misleading, did at least identify real entities and activities that could be to some degree measured and controlled. The HOTCO model seems to be designed with different purposes in mind. It is a model which can be used to simulate certain aspects of cognitive phenomena. Psychological research (and personal experience) suggest that emotional valences often affect cognitive judgments, and the HOTCO model creates analogous effects in its toy domain. Variants of the HOTCO model represent some of the dynamical features of human decision-making phenomena that are not represented in models without emotional valences (Thagard and Nerb 2002).

The HOTCO model is at least a rough model of certain kinds of cognitive phenomena, but it may not be anything more. It is, however, suggestive of a mechanism, so it is possibly a how-roughly model. It is possible to construe a HOTCO model to be similar to actual human cognitive systems in certain respects. The theoretical hypothesis suggests that the kind of mechanism that humans use for coming to decisions is one that uses some kind of coherence algorithm, and additionally that that algorithm will utilize both cognitive and affective links. The model does not identify the actual structures in the brain that would engage in these activities, and for this reason, among others, its warrant is quite limited. Nonetheless, interpreted this way, the model suggests something about *how* human decision-making processes work, and contrasts with other accounts of decision-making like the von Neumann and Morgenstern approach.

I say that this is only *possibly* a how-roughly explanation to emphasize how limited and tentative the data is that is available to support the theoretical claims of the model. Much of the value of such models is heuristic, in the sense that they suggest further lines of research. If, though, the theoretical hypotheses turn out to be true, these simple models will have identified some important features of the kinds of mechanisms that humans use to make decisions.

It will be helpful to contrast the HOTCO models with another model, GAGE, which Thagard claims is a more neurologically realistic model of emotional decision. I quote the abstract of Wagar and Thagard's paper on the GAGE model:

The authors present a neurological theory of how cognitive information and emotional information are integrated in the nucleus accumbens during effective decision making. They describe how the nucleus accumbens acts as a gateway to integrate cognitive information from the ventromedial prefrontal cortex and the hippocampus with emotional information from the amygdala. The authors have modeled this integration by a network of spiking artificial neurons organized into separate areas and used this computational model to simulate 2 kinds of cognitive–affective integration. The model simulates successful performance by people with normal cognitive–affective integration. The model also simulates the historical case of Phineas Gage as well as subsequent patients whose ability to make decisions became impeded by damage to the ventromedial prefrontal cortex.   (Wagar and Thagard 2004, 67)

The GAGE model and related experiments are meant to support and extend the work of Damasio's theories of affective cognition (Damasio 1994). Damasio's research begins with the famous case of Phineas Gage, a railroad foreman who suffered a serious brain injury when a tamping iron was shot through his scull as a result of an accidental explosion. That injury seriously damaged Gage's ventromedial prefrontal cortex (VMPFC), but despite this, Gage made a remarkable physical and cognitive recovery.

He is reported to have been able to read, write, speak, and perform many normal cognitive functions. However, brain damage appears to have had a substantial impact on his personality and in particular on his capacity to make prudent decisions. The Gage case is particularly well known, but it is part of a body of neurological evidence suggesting the importance of emotion for cognition.

As the abstract quoted above indicates, the GAGE model is intended to provide support for a more detailed hypothesis about how various brain regions control emotional cognition. It is clearly meant to be a mechanistic model, and it bears the evident markers. Entities, activities, and organization in the model are meant to correspond to entities, activities, and organization in the target, as is illustrated in the GAGE model schematic in Figure 3.3.

The entities in the diagram refer to actual parts of the brain, and the authors cite a variety of empirical studies that give evidence that these parts of the brain are in fact implicated in the decision-making process. The second way in which the GAGE model is supposed to be realistic is in its modeling of individual neurons. The GAGE model uses so-called artificial spiking neurons. In contrast to the artificial neurons in HOTCO and other basic connectionist networks, spiking neurons simulate variations in rate and duration of firing that are characteristic of biological neurons. These spikes travel with a certain frequency toward other neurons when threshold triggering conditions are met.

While Wagar and Thagard repeatedly describe this model as "much more realistic" than the HOTCO models, it is still very far from describing human decision-making mechanisms—both in terms of the behavior the model produces and in the mechanism that the model uses to produce the behavior. GAGE does not have the cognitive capacities to simulate an actual human being, even an emotionally damaged one. What the model can do is perform certain highly constrained experimental tasks in ways that are analogous to human agents. In their paper, Wagar and Thagard report on two experiments in which emotional responses appear to affect decision-making, and in which GAGE exhibits some features of human-like performances. I will briefly describe just one of these, the Iowa gambling task.

The Iowa gambling task is an experimental task in which subjects are asked to pick cards from one of four decks, which always offer a reward, and sometimes offer a

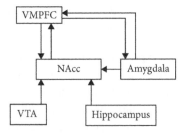

Figure 3.3 Schematic for the GAGE model, redrawn from Wagar and Thagard (2004). VMPFC = ventromedial prefrontal cortex; NAcc = nucleus accumbens; VTA = ventral tegmental area.

penalty. There are two "good decks," which have an overall payout that is better than the two "bad decks," but the bad decks are characterized by larger initial rewards. Human subjects engaging in the Iowa gambling task will typically figure out which are the good decks and will pursue a strategy of drawing from them. However, subjects with a damaged VMPFC will not learn and will pick from the bad decks. Wagar and Thagard hypothesize that "this is because such patients appear to be guided by immediate rewards (the high initial reward from the bad decks) rather than future outcomes (the overall profit from the good decks)" (2004, 73–4). They aimed to replicate this behavior using GAGE. They first performed the task with a "healthy" GAGE, and then repeated the task with a GAGE in which the simulated VMPFC was disconnected from the GAGE model. As with human subjects, the healthy GAGE was able to master the task, while the lesioned GAGE failed to do so.[16]

So much for the phenomena that the GAGE model produces. How does it do it? Here is part of Wagar and Thagard's description:

The model has 700 spiking neurons with 670 connections. The modeled regions consist of the VMPFC, the amygdala, the NAcc, the hippocampus, and the ventral tegmental area. Each region contains 100 neurons that receive input from other regions and/or external input and pass information on to other regions as well as 40 inhibitory interneurons. The pattern of afferent, efferent, and internal connectivity follows that of the proposed neural mechanism. The model includes intraregional connections and interregional connections....In constructing the model, we put emphasis on the incorporation of realistic network properties for the relative proportions of the various sets of connections composing the inter- and intraregional circuitry. Specific patterns of connectivity were classified as either sparse or dense on the basis of anatomical data.   (Wagar and Thagard 2004, 73)

While GAGE may be more realistic than HOTCO models, it is still a very long way from a complete description of the mechanism we find in human targets. Most obviously, there are not remotely enough neurons. The number of the neurons runs to the tens of billions or more, and while only a small fraction of those might be involved in the specific sorts of cognitive tasks under study, there will be orders of magnitude more than the 700 neurons and 670 connections in the model. The spiking neurons, too, while better perhaps than the HOTCO neurons, behave very differently than actual neurons, and the pattern of neuron connectivity with which the model starts is based on simple assumptions that are certainly false.

My aim in making these points is not to disparage this model, but to show that even a model whose construction is informed by detailed empirical research and which purports to characterize features of an actual mechanism can be highly idealized and unrealistic. The relation between this model and its target phenomena is nothing like

---

[16] Even this claim is true only in an attenuated sense that requires much interpretation. "Performing the task" consists of feeding activation vectors as inputs into the model, recording spiking patterns in the simulation, and then using signal analysis to detect "emotional reactions" from clusters of neurons. GAGE doesn't play the game the way that human subjects play the game.

the relation between my gear diagrams and the gears on my bike. We can still, however, identify degrees and respects in which this model is similar to actual targets. For instance, it has regions (i.e., bundles of neurons) that correspond to brain regions and it has strength and direction of connections between regions that are intended to be similar to those in actual brains. But these similarities do not amount to anything like a structural isomorphism between a model and its targets.

Still, one might think that we can hope to move toward models that are far more complete. Neuroscience is still in its infancy, but already much progress has been made in the nearly two decades since the HOTCO models were developed. In particular, developments in computer technology have allowed researchers to build large-scale brain models that may have as many neurons as human brains. But even in these cases, there are inevitable trade-offs. Some models model only certain regions of the brain, but in a great deal of detail at the neuronal level, while others have greater numbers of more simplified neurons.[17]

One certainly can find areas where we have much fuller mechanistic understanding than we do of human decision-making or, more generally, the large-scale structure and function of the human brain. For instance, mechanistic models of the action potential get closer to complete mechanisms than GAGE does, and we should have far more confidence in the explanatory import of these models than we do of models like GAGE. Yet even these models—which represent signal successes for neuroscience—are far from complete. For instance, models do not (and should not) seek to identify individual ions or ion channels, even though these are the parts of the mechanism whose activities and interactions are responsible for the phenomena. If you expect mechanistic models to faithfully represent all the entities and activities, then we have no mechanistic models in most of science. In this respect, consideration of models of simple machines like bicycles can give a very misleading picture of how scientific models represent.

The trade-offs in model-building in part reflect our technological and cognitive limitations, but this is not the only reason for them. They have also to do with the diversity of explanatory and other purposes to which we put models. As we have already noted, generality in mechanistic models is typically bought at the expense of mechanistic detail. Because, for instance, our brains are different from monkey brains, and indeed my brain is different from your brain, more detailed brain models will apply to fewer targets. Additionally, in the complex systems that are of interest to us, there are many different phenomena that occur at many different levels, and which generalize differently across systems. For instance, we may have one model of neurological phenomena that applies broadly across a range of species, while other models account for species-specific or even finer-grained phenomena.

---

[17] See Eliasmith and Trujillo (2014) for a review of some recent large-scale brain models. Eliasmith and Trujillo argue for a kind of many-models pluralism in brain research that is consistent with the account of modeling I am developing here.

In the context of psychological and neuroscientific phenomena, both Hochstein (2015) and Stinson (2016) have made the case for the many-models model. Stinson argues that cognitive models cannot always be "seamlessly" filled in with mechanistic detail at the neurological level, while Hochstein suggests that explanations are inevitably distributed across a variety of models that describe different features of a mechanism, and its many levels of entities, activities, and organization. I suggest that the considerations that Hochstein and Stinson have identified that support a many-models approach will apply quite generally. Ultimately, the need for many models of mechanisms derives from the variety of purposes we have in modeling, the plurality of phenomena with a complex hierarchy of mechanisms, and the inevitable abstractions and idealizations that help us find generality in a world of mechanisms that are ultimately particular, localized, and heterogeneous.

What we can conclude from reflection on these examples is that the perfect-model model is itself a bad idealization of science. Models are always partial and incomplete, and accordingly a realistic account of explanatory norms should seek to understand how partial and incomplete models can explain, rather than treating them as imperfect approximations of an ideal model.

3.7 Mechanistic and Narrative Explanation

Mechanistic explanations, as Bechtel and Abrahamsen (2005) put it, "explain why by explaining how." But this can equally be said of narrative explanations. Because narrative explanations are typically understood to occur within the historical sciences (and especially the study of human history) while mechanistic explanations are typically employed in the natural sciences, one might think that these are very different kinds of how explanations. I will argue, however, that they are not. On the one hand, paradigms of mechanistic explanation contain narrative elements, and on the other, narratives are appropriately understood as models of mechanisms.

On the connection between narratives and mechanisms, Daniel Little writes:

> The idea of causal mechanisms fits very well into a characteristic mode of historical writing, the form of a causal narrative. Essentially the idea is that a causal narrative of a complicated outcome or occurrence is an orderly analysis of the sequence of events and the causal processes that connected them, leading from a set of initial conditions to the outcome in question. The narrative pulls together our best understanding of the causal relations, mechanisms, and conditions that were involved in the process and arranges them in an appropriate temporal order. It is a series of answers to "why and how did X occur?" designed to give us an understanding of the full unfolding of the process.   (Little 2018, 417)

On Little's account, a narrative looks to be a description of what I have called a process mechanism. It describes sequences of events (which will typically be entities acting and interacting), and shows how their arrangement in space and time brought about some outcome—the end state of the mechanism from some set of start-up or initial conditions.

What appears most to distinguish narrative from typical mechanistic explanations is the particularity of the explanandum. Historical narratives are typically offered to explain how singular events occurred. What led to the fall of the Roman Empire or to the passage of women's suffrage laws in England? The particular events and the processes that led to them are not repeatable. The story of the fall of the Roman Empire is a different story than the story of the fall of the Hapsburg Empire, and while historians may rightly point to similarities between these stories, in the singular causal explanation of historical events the particularities take center stage. This stands in stark contrast to the examples of mechanistic explanations that I have discussed in this chapter. Explanations of the action potential, human decision-making, or bicycle gearing are explanations of general phenomena. Action potentials, for instance, are triggered countless times in countless cells in countless organisms, and the story is similar most every time.

But the fact that mechanistic explanations of the action potential and other recurrent phenomena are general does not detract from the narrative character of those explanations. There still is a story to be told. Bechtel and Abrahamsen (2005) are explicit about this. They emphasize that mechanistic explanations involve simulations—running through the various stages in the processes that lead to the end-state of the mechanism. Sometimes these simulations are mental, with a scientist working through each of the steps, and seeing how things go. This is a kind of storytelling about the mechanistic process. Frequently, though, the processes are too complicated for mental simulation. Proper understanding of how the process mechanism gives rise to the phenomenon requires a computational model to simulate the complex interactions of the mechanism's parts over time (Bechtel and Abrahamsen 2010).

Bechtel and Abrahamsen are not the only ones to have connected simulation and narration. Wise (2011) has argued that there are significant narrative elements in explanations in physics—citing as two examples models of quantum chaos and of snowflake formation. For Wise, narrative explanations in science explain phenomena, not by subsuming them under general laws, but by "growing them." Wise sees the expanded use of simulations in the physical sciences as providing a new kind of narrative technology. They allow scientists to tell historical stories of very complex processes. Something similar seems to be going on with the growth of agent- or individual-based models that are now frequently being used in some parts of biology and in various social sciences (DeAngelis and Mooij 2005; Hedström and Ylikoski 2010; Little 2011).

Narrative explanations appear at first glance to emphasize the causal over the constitutive aspect of mechanistic explanations. Narrative explanations tell how one thing leads to another—the various links in causal chains whereby the entities involved in the narrative act to produce some outcome. Undoubtedly this is true for certain kinds of narrative explanations. The story of how I came to marry my wife is a narrative that describes the events leading to our meeting, and how one thing led to another until we got married. But most narratives of the sort constructed by historians contain not just causal but constitutive elements. Narratives describe not just how the activities and interactions of individuals produce changes to those individuals. They also describe how those individual activities and interactions constitute larger interactions and events. Consider a narrative account of a battle. How does one explain the battle's outcome? Typically one refers to actions of individual people within the battle, or perhaps larger aggregates like the battalions, regiments, or other units that make up the armies. At Gettysburg, the Confederate army lost the battle in part because Lee ordered Pickett to charge at the well-fortified Union lines, and Pickett and his men charged. There is causation here, for instance Lee's issuing of orders causing Pickett to issue further orders, but there is also constitution, where what happened to Lee's army as a whole depended constitutively on what happened to its parts. Lee's army suffered heavy casualties and became demoralized because a part of it (Pickett's division) suffered heavy casualties and became demoralized, and this happened to Pickett's division because it happened to the parts of Pickett's division—the brigades, battalions, and ultimately men, who constituted the division. The changing of scales within narrative explanations from aggregates to their parts and back is a crucial feature of explanations in history and the social sciences, and is central to successful narrative explanation (Gaddis 2002; Little 2006).

What this all suggests is that narrative explanation is a kind of mechanistic explanation, and the explanations we more typically think of as mechanistic involve narratives. In both cases, the explanation involves constructing a model (for a narrative is a model) that characterizes the mechanism responsible for the phenomenon to be explained. Narratives, like other mechanistic models, are used for different purposes, and these different purposes and other pragmatic considerations will guide the abstractions and idealizations used in the model. This in turn implies that there is not one master narrative, but many—narratives that pick out different causally relevant aspects of the entities and their activities that are responsible for the explanandum.

The chief difference between historical and scientific narratives is not one of kind, but of degree. Historians (both of the human and natural worlds) are concerned more with the explanations of singular events, and there is often a higher degree of contingency in how the parts of these historical mechanisms come together to bring about these events (Glennan 2010b; Glennan 2014). Scientists, conversely, tend to be more concerned with more robust mechanisms that give rise to more regular phenomena. But these are only tendencies, and counterexamples are easy to find.

3.8 Conclusion

In this chapter I have moved away from ontological questions about the nature of mechanisms to epistemic and pragmatic questions about how to represent mechanisms with models, and how to use those models to explain. Because our purposes and epistemic constraints are various, the models that can rightfully be employed to represent a mechanism and explain some phenomenon will be many. Different models will be similar to target mechanisms in different degrees and respects. There is no perfect model of a mechanism.

Though mechanisms are particulars, the models scientists use to represent them can often be employed to represent a whole class of these mechanisms. This is the basic source of generality. But for mechanistic models to represent a larger class of mechanisms, those models must employ abstraction and idealization. Whether one mechanism is like another depends upon what features you are looking at and how closely you look.

The account of models offered here will be key to the topics to which I shall now turn. In the next two chapters I will explore how we classify entities, activities, and mechanisms into kinds, and I shall argue that we cannot understand kinds apart from the models used to represent those kinds. I shall also return to the relation between mechanistic and other kinds of models in Chapter 8, where I compare the ways in which mechanistic and other kinds of models explain.

4

Mechanisms, Models, and Kinds

Scientists and ordinary folk divide things in the world into kinds. There are kinds of things (or entities)—kinds of rocks, plants, celestial bodies, and bicycles. There are also kinds of activities or processes—kinds of chemical reactions, kinds of cooking, kinds of communication, kinds of dancing. There are also kinds of mechanisms. Our need for kinds is an expression of our craving for generality, or as Goodman (1955) or Quine (1969) would put it, to ground inductions. We can say of kinds of things that they tend to have similar properties and dispositions, and that they tend to act in the same kind of ways. Metals can conduct electricity, oaks produce acorns, males and females of many species may form pair bonds, consumers will tend to buy products at the lowest proffered price. Categorizing an entity, an activity, or a mechanism as an instance of a kind allows us to understand, predict, and control the things we categorize. Our efforts will be of only limited success, as these generalizations will be gappy, context-sensitive, and hedged. Not all oaks produce acorns, and certainly not all acorns oaks. Sexually dimorphic animals do not always form pair bonds, and when they do they do not always pair with the opposite sex, and indeed there is a lot of vagueness in what might count as pair-bonding in the first place. Consumers certainly do not always buy from the cheapest seller.

In this chapter I will explore how we use kind terms to characterize mechanisms and their constituents. The task has both descriptive and metaphysical aspects. In the first instance, we want to observe principles and techniques for sorting mechanisms, and the parts and activities that make them up, into kinds. Secondly, we need some understanding of the ontological status of kinds. For instance, are kinds constructed or discovered, and in what sense is there a reality to kinds apart from their instances?

There are a variety of kinds of kinds. Brian Ellis (2001, 74) has argued that an adequate account of natural kinds should recognize three primary categories: substantive kinds, dynamic kinds, and property kinds. Substantive kinds are the most familiar, including "stuff" like water or gold, and "things" like cats or cars. The stuff/thing distinction mirrors the grammatical distinction between mass and count nouns, and a semantic distinction between non-sortal and sortal concepts. You can count things like cats and cars (using sortal concepts), while you can measure things like water or gold (using non-sortal concepts). The entities or parts of which mechanisms are composed are substantive and fall under sortal concepts. The activities and interactions in mechanisms are instances of dynamic kinds—happenings like bonding, walking, or colliding.[1] Mechanisms themselves seem to fit into none of these categories—they consist of organized entities and activities—and their identity is connected to both their dynamic and substantive characteristics. Mechanism kind concepts, however, are sortals. You can count mechanisms.

I have argued that mechanisms are particulars located at particular places and times, and that different instances of mechanisms (or entities or activities) of the same kind will exhibit variation in their constituents, organization, and behavior. These facts will push us in the direction of nominalism about kinds. We cannot understand mechanism (or entity or activity) kinds without using models of these particulars, and the attendant strategies of abstraction and idealization discussed in Chapter 3. Because models are used for various purposes, pragmatic consideration will come into play as we use models to represent kinds of mechanisms.

An analogy might be helpful here. The biologist Ernst Mayr (1959) famously argued that Darwin's central achievement was the rejection of essentialism and its replacement with "population thinking." Population thinking in the Darwinian sense involves seeing the individual organism as real, recognizing the existence of variation across individuals, and seeing ideal types as epiphenomenal abstractions from these individuals. What I am advocating is a kind of population thinking about mechanisms. The real is the individual and properties of the type are either abstractions and idealizations from or statistical averages over features of individuals.[2]

This is not to say that we can do without kinds. While mechanisms are particulars that are constituted by other particulars (i.e., by particular entities or activities), the particulars that constitute them are not bare particulars, but are particular instances of kinds of things. So a particular watch is made of particular springs and gears, but these particulars *are* springs and gears, and in characterizing them as such we are characterizing them as instances of kinds. Any representation of or reference to a particular is thus inevitably shaped by our way of dividing particulars into kinds, and thus, even if reality is particular, our way of making particulars intelligible cannot fully honor this fact. Another way of putting this is that the way in which we divide the world into parts is inevitably influenced by our understanding of kinds of parts.

I will call the general approach to kinds that I offer in this chapter a "models-first strategy." It suggests that entities are of the same kind when the same model can be used to represent them. But before fleshing out this strategy, I will review a few standard approaches to thinking about kinds in the philosophical literature, so that we can better understand the models-first strategy by way of contrast.

---

[1] It is possible to distinguish sortal and non-sortal activity concepts, though the boundary is vague. Some activities and interactions can clearly be counted—hitting something, forming or breaking of chemical bonds. Other activities—sweating, dancing—cannot so easily be counted, and there are intermediate cases. Activity sortals are possible where there are clear activity boundaries (see Section 2.5).

[2] There is more to population thinking in evolutionary biology than this anti-essentialism. Population thinking has a special role because of the fact that variation is necessary for natural selection. The exact nature and consequences of population thinking has been a subject of continuing discussion since Mayr's original formulation (Ariew 2008).

4.1 How Real and Natural Are Natural Kinds?

Discussions of natural kinds are often connected with realism. Realists of most stripes believe that there are kinds of things—electrons, elements, stars—and that such kinds of things exist independently of our knowledge of them, and indeed independently of our existence. Instrumentalists and constructivists insist that our ways of classifying the world reflect our interests and purposes. The extreme positions are, in my view, implausible and uninteresting; the challenge is to sort out a middle ground which respects both the constructive role of individuals and institutions in classifying the world and the very real constraints that that world puts on successful classifications.

Bird and Tobin (2016) make a distinction between two kinds of realism about natural kinds. *Weak realism*, which they also refer to as naturalism about kinds, holds that there are natural classifications in the world, natural in the sense of being mind-independent, non-arbitrary, and not reflecting human interests. A weak realist can hold that kind terms like "carbon," "water," "giraffe," or "galaxy" refer to such natural classifications. The weak realist grants that some kind terms may not represent kinds at all. They may doubt that there is a mind-independent classification of entities into colors or genders or races. The weak realist's commitment to the naturalness of some classifications means that they think that it is an important goal of science to find the real kinds and discard the ersatz ones.

What Bird and Tobin call *strong realism* is a view about kinds as abstract entities. The strong realist asserts not merely that there are natural ways to classify particulars into kinds, say classifying chunks of ore as gold or silver, but that kinds like gold and silver are themselves real things (Bird and Tobin 2016). This sort of realism sits uncomfortably with the particularism of the New Mechanist's ontology, but fortunately weak realism is all the realism needed to underwrite the account of natural classification I will offer here.

What is at stake in calling a kind "natural"? Commonly, natural kinds are distinguished from social or human kinds (Khalidi 2013). The simplest way to make this distinction is to think of natural kinds as the referents of kind terms in the natural sciences, while social or human kinds are referents of kind terms from the social sciences; so, for instance, *magnet* is a natural kind while *ritual* is a social kind. There appear to be some interesting differences between natural kinds and at least many social kinds. Most notably, as Hacking has argued, certain social kinds appear to be "interactive" in the sense that human beliefs about the kinds change the kinds in various ways (Hacking 1986; 1999). Notwithstanding the differences, I shall treat both natural and social kinds as natural kinds in a broader sense. The basic reason is that the social world is part of the natural world.[3] Human beings of course have some interesting

---

[3] I am here taking sides in an old debate about the relation between the natural and the social sciences—one with roots in the birth of the social sciences in the nineteenth century. A classic formulation of the debate can be found in the exchange between Charles Taylor (1971) and Thomas Kuhn (1991). I concur with Kuhn that part of the reason that Taylor and others have mistakenly tried to separate the human from

properties (like the capacity to form representations, to communicate, and to cooperate), but there are not clear lines between the human and the non-human here. Many animals have very human-like cognitive and affective capacities, as well as some similar forms of social organization—and basic capacities to represent are widespread in biological systems.

Equally important, we do not want to conflate the question of whether a kind is natural or social with the question of whether a particular classification is mind-independent or not. If the weak realist in the sense of Bird and Tobin is correct, then it is possible for both natural and social kind terms to pick out genuine kinds. In saying that social kinds may be "mind-independent" we must be careful to distinguish two senses in which a kind can be mind-independent. The trivial sense in which most or all social kinds are not mind-independent is this: social kinds are kinds that exist only in the context of groups of individuals that have minds. If there are no minds, there is no money, there are no patriarchal or matriarchal societies, and there are no rituals; certainly, there is no mental illness without minds! But to say this is very different from saying that these kinds are mind-dependent in the sense that they are constructs of our theorizing. Presumably elephant troops were matriarchal long before any scientists invented concepts like "matriarchy" to describe these structures. One can believe that a certain kind of thing or phenomenon arises only in the presence of one or more cognitive agents while still believing that the phenomenon or thing is mind-independent, in the sense that it exists independently of the theorizing of scientific observers.

Kind terms group particulars into classes or sets. To call something a eukaryote or a Chevrolet is in part to say that it is a member of a set along with the other eukaryotes or Chevrolets. Looking at kind terms in this way, we can see that the philosophical challenge is to describe the criteria for set membership. The most traditional view, often associated with Locke, is that natural kinds are defined by their real essences. Real essences would be intrinsic properties of a thing that make it the kind of thing that it is. All and only instances of the kind will have this real essence. The atomic elements provide a paradigmatic example of the sorts of kinds for which the real essence view is plausible. At least part of the real essence of any chemical element is its atomic number—the number of protons in its nucleus. To be a carbon atom is to have six protons in the nucleus—no more, no less. The number of protons in the nucleus is an intrinsic property of an atom, and its chemical properties and dispositions will follow from its intrinsic properties.[4] This is clearly a realist view of kinds, where kinds are discovered, not constructed.

the natural sciences has to do with a false presumption that, unlike the human sciences, the natural sciences are free of "interpretive" or value-laden elements.

[4] This is not to say that all of the properties of an element are fixed by atomic number alone. For instance, the atomic number fixes the charge of the nucleus, but not its weight, which depends also upon the number of neutrons, which may vary among isotopes of the same element. This is not a problem for the concept of a real essence, as kinds may be divided into lower taxa. For instance, carbon is defined by atomic number, but isotopes of carbon, e.g., C_{12} and C_{14}, are distinct kinds defined by distinct mass numbers.

The real-essence approach may work for atomic elements, but there are many sorts of scientifically interesting kinds where it is not possible to identify a set of intrinsic properties necessary and sufficient for kind membership. The next best thing might be a cluster analysis, where things are classified into kinds in virtue of sharing many or most of a cluster of properties. On cluster analyses kinds may have fuzzy boundaries. Take as an example biological sex in human beings. There are a cluster of properties that distinguish males from females—chromosomal, hormonal, anatomical, and physiological. Most human beings will exemplify the whole cluster of properties characteristic with either males or females, but occasionally there will be ambiguous or intersex cases. In other cases, e.g., in the distinction between juveniles and adults, there may be paradigm cases but also a wide range of intermediate forms.

The difficulty with cluster concepts lies not so much in their vagueness as in the question as to which clusters of properties should be taken to form kinds. There may for instance be a cluster of traits that characterize dogs as a kind, but there are other clusters of properties (e.g., of soft, fuzzy, cotton, square-shaped things) which do not intuitively form a kind. It is largely to address this sort of problem that Boyd introduced his influential homeostatic property cluster (HPC) conception of kinds. Boyd (1989; 1991) argues that natural kinds are not merely conventionally or conceptually related clusters of properties, but clusters of properties whose co-occurrence is somehow explained or produced by homeostatic mechanisms. "Homeostatic" in this sense means similarity-generating. The idea is that in a genuine natural kind, there is some sort of mechanism which keeps the cluster of properties together.

Boyd's HPC approach is motivated by the desire to make sense of biological kinds, and in particular of species as kinds. While there is no set of properties shared by all and only those individuals that are members of a species, there are clusters of properties—mostly shared by most members—and there are mechanisms that produce this clustering. Most tigers, for instance, have tails, stripes, big teeth, and a host of other traits that are produced via developmental mechanisms relying upon genes and environment, and there are mechanisms both internal (reproductive isolation) and external (selective pressure) which tend to maintain this cluster of properties.

One challenge for the HPC conception of kind is that it seems to require an antecedent notion of a kind of mechanism. For entities to belong to a kind, it seems appropriate to require that the similarity-generating mechanisms for the property cluster be of the same kind. As Craver points out in a discussion of the HPC approach, it is commonplace to engage in "taxonomic reform" both by lumping together heterogeneous phenomena (i.e., patterns or property clusters) when they are generated by the same kind of mechanism, and by splitting kinds with similar property clusters when the clusters are generated by different kinds of mechanisms. As Craver puts it:

The splitting strategy can be expressed informally in the directive: if you find that a single cluster of properties is explained by more than one mechanism, split the cluster into subset clusters, each of which is explained by a single mechanism.... The lumping strategy is expressed in the informal directive: if you find that two or more putatively distinct kinds are explained by the same mechanism, lump the putative kinds into one.   (Craver 2009, 581)

As an example of splitting, consider the classification of diseases. Different diseases can produce similar clusters of symptoms, but we do not identify the disease with the symptoms, but rather with the agent or mechanism that produces the symptoms. In lumping we have the converse, as when we lump together phenomenally distinct clusters of properties, for instance rusting and burning, via a similar set of productive mechanisms (redox reactions). The centrality of lumping and splitting as taxonomic principles suggests that the heart of an HPC kind is the similarity-generating mechanisms, with it being those mechanisms rather than the properties they cluster that make the kind. Accordingly, for the HPC approach to work as a method for classifying things like organisms or diseases, one must have an independent account of how one should classify mechanisms into kinds. To this we now turn.

4.2 Mechanism Kinds and the Models-First Approach

To consider the variety of ways in which particular mechanism tokens or instances can be classified into types and kinds, it is helpful to return to the definition of minimal mechanism from Chapter 2:

A mechanism for a phenomenon consists of entities (or parts) whose activities and interactions are organized so as to be responsible for the phenomenon.

If two or more mechanism tokens are instances of a kind in virtue of the existence of similarities between them, minimal mechanism gives us a map for the kinds of similarities that might unite those tokens into kinds. Two mechanisms could be of the same kind in virtue of having similar organization, or being made of similar constituents. Minimal mechanism suggests that similarities can be evaluated over the following dimensions:

1. Kinds of phenomena
2. Kinds of entities (parts)
3. Kinds of activities (including interactions)
4. Kinds of organization.

Beyond these, another important way of delineating kinds of mechanism is by considering how they came to be. I shall call this dimension:

5. Kinds of etiology.

The strongest criterion for classifying two mechanism instances as being of the same kind would require qualitative identity with respect to all five features.[5] But this criterion

---

[5] Qualitative identity means simply identity with respect to properties, as opposed to numerical identity, which is a relation that holds only between a thing and itself. I am assimilating the type-token distinction

is absurdly strong. On this criterion, each mechanism would literally be one of a kind. The absurdly strong criterion can be loosened in one of two ways. First, one can loosen the requirement of qualitative identity to qualitative similarity, so, e.g., two mechanisms are of the same kind with respect to the phenomena they produce if the token phenomena are sufficiently similar. Second, one can loosen the requirement that mechanism tokens must be identical with respect to all five of the features. This would mean that we might count two particular mechanisms as being of the same kind if they produced the same kind of phenomena, even if the entities, activities, and organization that produce these phenomena differ across the particulars. If this position is right, there are lots of kinds that will create cross-cutting classifications. It will not be the case that a mechanism instance is an instance of just one kind, or that it fits in a unique place within a hierarchy of kinds.

The position I am suggesting here falls within a family of philosophical positions which have sought at once to acknowledge the essential role of perspectives or stances within the complex sciences while taking seriously the reality or objectivity of entities and kinds seen from these perspectives and stances. Examples include Wimsatt's (1994) account of perspectives and rainforest ontology, Dennett's (1991) account of real patterns, Dupre's (1993) promiscuous realism, Giere's (2004) perspectival realism, Mitchell's (2009) integrative pluralism, Kellert, Longino, and Waters's (2006) scientific pluralism, and Ladyman and Ross's (2007) rainforest realism. Despite their differences, all of them share what Mitchell calls a "pluralist-realist approach to ontology, which suggests not that there are multiple worlds, but that there are multiple correct ways to parse our world, individuating a variety of objects and processes that reflect both causal structures and our interests" (Mitchell 2009, 13). I think that collectively these and other studies make the case for some sort of pluralist-realist kind of view.[6]

In Chapter 3, I argued that generality in representation of mechanisms is achieved by the use of abstract models. An abstract model may be used to represent a variety of mechanism instances, insofar as those instances resemble the model in specifiable degrees and respects. We can say then that, to the extent that two particular mechanism instances are representable by the same model, they are instances of the same kind. This is the crux of the approach I call the models-first approach to mechanism kinds.

The models-first approach leads naturally to a pluralist and cross-classifying approach to natural kinds. This is because a pair of targets might be representable by a common model with respect to one feature (say, kind of organization) while being classified with other quite different targets via different models that represent similarities

and the kind-instance distinction. This is fairly common in the literature, and will not do any harm so long as one is not focusing on metaphysical positions of the sort I described as "strong realism" above.

[6] Critics may worry that we should be cautious to read ontology too quickly out of our practices. Alan Love (2008), for instance, makes a similarly pluralist case for typologies in biology, motivated by what he calls an examination of the "tactics" of biology—the specific tools biologists use in various contexts to pursue their epistemic goals. Love argues that this typological pluralism can be endorsed without embracing any particular metaphysical commitments about kinds. For an empiricist alternative motivated by similar concerns about perspective see van Fraassen (2008).

in other dimensions (say, kind of etiology or kinds of entities). For example, we can characterize two mechanism instances as being negative feedback mechanisms, because we can represent their organization using the same model—a model indicating a looping relation between two nodes in a system. The same basic model might be used to represent temperature regulation in a home heating and cooling system or hormonal regulation in an endocrine system. While in this respect (organization), two mechanism instances would be of similar kinds—in other respects, say in their etiology or the kinds of entities and activities that constitute the mechanisms, they could be quite different. And even in a single typological dimension, such as organization, whether two instances will count as belonging to the same kind will depend upon the grain of the model and the construal. At its coarsest, negative feedback mechanisms are any mechanism that can be represented with two quantitative variables where one is positively linked to the second, and the second negatively linked to the first. Two mechanisms might be of this kind, at this very course level of description, but have quite different organizations as one develops more detailed models.

Because the relation of model to target is one of similarity, whether natural kinds on the models-first view can be thought to be real and objective (in the weak sense) will depend upon whether there are objective degrees of similarity between various targets and their models. And here there are prima facie reasons for concern. Similarity is often thought, like beauty, to be in the eye of the beholder, and hence not a ground for dividing the world into real kinds.[7]

In defense of similarity we can appeal to an observation made by John Heil (2003, sec. 14.2). Much of the worry about similarity comes from conflating object similarity and property similarity. On the one hand, there do not appear to be objective levels of similarity among objects. Given any two objects, we can typically find indefinitely many ways in which they are similar or dissimilar. Looking around at my office here there are two things I immediately see: a model of the giant Ferris wheel in Vienna's Prater and a coffee cup I got last year at a university function. They have much in common—both have circular shapes, gold lettering, and they have similar weights. Clearly that game can be played with any two objects, so the existence of similarities like this will not get us mind-independent kinds. But as Heil points out, if we restrict ourselves to similarities between properties, these concerns about subjectivity recede: If I pick a property like weight, I can find any set of objects in my office, and get objective measures of degree of similarity. The coffee cup and the telephone are very similar to each other in weight, while the pencil and the pen are similar to each other in shape.

---

[7] My focus here is on the practical problem of understanding the grounds of comparative similarity judgments. I shall not explore some of the more basic metaphysical and theoretical issues surrounding similarity, a notion that some prominent philosophers, like Goodman and Quine, have found deeply suspicious. Quine (1969) calls similarity "scientifically disreputable"—but he thinks that many categories that most contemporary philosophers of science and metaphysicians take for granted—like natural kinds, properties, dispositions, and causes—are disreputable as well. I will not try to refight this battle here. For a discussion of the overall prospects for similarity in light of criticisms raised by Goodman and Quine see Decock and Douven (2011).

Applying Heil's point to mechanisms, we can say that there is no way to say of one mechanism overall that it is objectively more or less similar to another mechanism. But once we attend to particular dimensions of similarity, objective judgments become possible. Consider again a car engine, understood as a mechanism responsible for producing rotation of a drive shaft. There is no objective ordering of car engines by overall similarity. Still, we can objectively measure aspects of similarity by identifying features of the engine's parts, activities, and organization—for instance, diesel versus gasoline, number of pistons, fuel injectors versus carburetors, turbocharged or not—as well as aspects of the engine's behavior (i.e., of the phenomena produced by the mechanism) like fuel efficiency, horsepower, or torque curve. These various features lead to a proliferation of kinds of engines—V8s, diesels, turbocharged engines, high-efficiency engines, and so on. These kinds are perfectly real in the weak sense of being objective and mind-independent features of the world. One does not need to believe that there are abstract entities corresponding to each of these engine types. It is enough that these objective similarities can be truth-makers for claims that a set of particulars can be represented by a common model.

There may be times when one wishes to make overall similarity comparisons—to ask not just to what degree a model is similar to a target in some respect, but to try to weight the various degrees and respects. Such comparisons would allow one to classify the overall degree to which a model fitted a target, or which of two models was a better model, etc. Such judgments can be helpful so long as we recognize that the weights chosen, implicitly or explicitly, reflect epistemic and pragmatic circumstances and goals. Similarity is not just in the eye of the beholder, but it does depend upon the beholder's goals and interests.[8]

4.3 Kinds of Activities and Entities

One of the dimensions of similarity that we can use to classify mechanisms as being of a similar kind is similarity in the entities and activities that constitute them. In this section I want to explore this dimension and say more specifically how it is that we can classify entities or activities into kinds. I will not try to provide a substantive catalog of what kinds of entities and activities there are. The varieties of entities and activities that constitute mechanisms in nature is a subject for scientific investigation rather than philosophical theorizing. What we can do, and indeed need to do, is to explore a few examples to help understand something about the conceptual and ontological relationships that obtain between particular entities and activities and their kinds.

Let us first consider activities. In Chapter 3 I argued that abstraction is not a feature to be found in the world, but is rather a feature of our representations of the world. Any particular activity in the world will be fully concrete, though our representations of

---

[8] Weisberg (2013) develops such an account of the relation between models and targets, but he emphasizes the ways in which different weightings will correspond to different modeling goals.

that activity may be more or less abstract. But if particular activities are instances of kinds in the weakly realist sense we have discussed, there must be features that particular activities share that make them activities of the same kind.

My suggestion is that we can understand the nature of these features by borrowing a concept that has sometimes been used to characterize the relationship between more and less specific properties—the relation between determinates and determinables. The idea is that properties come in different degrees of specificity, and that having a determinate property is a specific way of having a determinable property. For instance, shape is a property that an object can have, but it is a rather non-specific one; there are various more specific properties (like roundness or squareness) that are determinates of the property shape. The determinate/determinable relation is a relative one, in that there can be more and less determinate properties. For instance, being curved is a determinate of being shaped, being elliptical is a determinate of being curved, and being circular is a determinate of being elliptical. Stephen Yablo, who applies this distinction in an attempt to understand the relation between physical and mental properties, puts the general distinction with respect to properties thus:

P determines Q iff: for a thing to be P is for it to be Q, not *simpliciter*, but in a specific way.
(Yablo 1992, 252)

Now activities are not properties or relations; they are things that an entity or entities do over some period of time. Nonetheless activities come in kinds that seem to stand in determinate/determinable relations. For instance, bonding is a kind of interaction, chemical bonding is a kind of bonding, and covalent bonding is a kind of chemical bonding. It is easy enough to offer a determination schema for activities analogous to Yablo's. For example, restricting ourselves to two-place interactions, we can say:

A_2 determines A_1 iff: for any entities x and y to do A_2 is for them to do A_1, not *simpliciter*, but in a specific way.

For instance, covalent bonding is a determinate of the determinable bonding. For atoms within a molecule to form a covalent bond is for them to bond, not *simpliciter*, but in a specific way. Activities, as we have said, are productive processes that connect one thing to another. There is not one thing that is production. Just as there is no object that has shape without having a specific shape, there is no activity that produces without producing in a specific way. If we stick to the level of generic production, we have little that we can say about what production is or how to find it.

More determinate kinds of activities require particular kinds of entities that can engage in them, particular mechanisms by which the activities are carried out, particular boundary conditions, particular inhibitors of production, etc. With increasing determinateness comes a lot of activity-specific features that determine how we can observe, measure, and control the activity. For instance, when the door to my room squeaks, we say more than that the pieces of the hinge interact. We say more specifically that the hinge pin and the two sides of the hinge rub against each other as the door moves. This rather specific form of interaction has a rather specific and annoying effect—the creak—and the kind of interaction determines the kinds of interventions that will change the effect. For instance, shining light on the hinge will not do anything, but spraying WD40 will get rid of the creak. I never interact with the hinge "generically"; I do something quite specific—like spraying WD40 on it.

Consider another example, the activity of seeing. Seeing can be represented as a two-place activity (i.e., an interaction) between a seer and a sight (some object or action or event that is seen). Rather broadly we can describe seeing as an interaction whereby light transmitted from the sight produces a representation of the sight in the seer. Sight interactions can only take place between appropriate kinds of entities. The seer must have mechanisms for detecting light reflected off the sight and for transforming it into representations, while the sight must be the sort of thing that influences the behavior of light in ways that can be detected by seers. Neither electrons nor elections can be seen, nor can they see. Even though seeing is a much more specific category than interaction, it is probably still far too indeterminate for us to clearly characterize its productive character. For that we must get more specific. There are many different kinds of seers and many different kinds of sights. Frogs see differently and see different things than do humans or flies. And according to this rather vague characterization of seeing, things without central nervous systems (like certain kinds of robots) might be said to see as well. The relation between seeing and interacting is one of determinate to determinable: seeing is interacting, but interacting in a specific way. With more determinate forms of interaction come clearer truth conditions. I do not, for instance, have clear truth conditions for whether Franz and Sisi interacted at the party, but I do have clear truth conditions for them seeing each other.

But what exactly makes an activity more or less determinate? Are there really more and less determinate activities, or is it only that there are more or less determinate descriptions of activities? In other words, is there indeterminateness in some activities themselves, or is the indeterminateness of activities just an indeterminateness in our representations of those activities? The models-first approach to kinds suggests the latter view. Any specific activity or interaction will be fully determinate. To be a kind of activity is to be suitably represented by a model of that kind of activity, and more abstract models will represent a broader class of particular activities, and will leave undetermined more features of those activities. This weakly realist view does not, though, prevent us from speaking truly of a hierarchy of activity kinds. There really are features that activities share that make them activities of a kind.[9]

The determination relation does not produce a branching hierarchy of activities. There is not a taxonomy of activities beginning at a root and proceeding in Linnaean

[9] In thinking of the determinate/determinable relation in this way I am taking what Wilson (2013) has called a semantic approach. This contrasts with a metaphysical approach of the sort advocated by, e.g., Yablo (1992); Rosen (2010); and Wilson (2013). Although I shall not explore the point, my hunch is that there is room for a mixed approach where some cases of indeterminateness are semantic, others epistemic, and other metaphysical.

fashion into genus and species. Concrete activities may simultaneously be determinates of several less determinate activities. Consider for instance the activity of talking. Talking is a determinate form of noise-making. It may also be a determinate form of other activities, like communicating, flirting, intimidating. A given concrete activity or interaction will be a determinate of many determinables. This is not a problem. Just as an object may be both blue and spherical, an activity may be both talking and flirting. The one does not exclude the other.

While I have said that any particular activity or interaction will be fully determinate, we must be careful how to interpret this claim. If we consider some particular activity—for instance, the Cubs' short-stop's catching of a ball to get the third out of the ninth inning in a game on a particular day in August 2015 at Wrigley field—this is a maximally specific happening that can be characterized by a model at any level of specificity one might desire. But this should not be taken to imply that there is a complete or perfect model that captures all features of the catch. That is the trap of the perfect-model model. Neither should it be taken to imply that the boundaries of the activity or the entities are fully determined. As we have seen from our discussion of boundaries in Chapter 2, there may not be a fully objective fact of the matter about where either an activity or entity begins or ends.

Much of what I have said about more and less determinate activities can be applied to entities (objects) as well. We can construct a similar schema:

Entity E_2 determines E_1 iff: an E_2 is an E_1, not *simpliciter*, but in a specific way.

For instance we can say that a screw is a determinate of a fastener, because to be a screw is to be a fastener, not *simpliciter*, but in a specific way.

Just as increasingly determinate activities require increasingly determinate entities, the converse holds. Fasteners fasten, somehow, while nails must be nailed and screws screwed. Second, there are hierarchies of increasingly determinate entities. For instance, screws are determinates of fasteners, but Phillips head screws are determinates of screws. Third, it is not the case that determination relations yield a single branching hierarchy of kinds of entities. A particular entity like a screw will be fully determinate, but it will be an instance of a variety of kinds of entities. A screw may simultaneously be a Phillips or flat head, a #4 or #6 screw, a 2 inch or 1 inch, a brass fastener or a steel fastener, etc.

As the example of screws and fasteners suggests, entities—especially manufactured objects—are often classified by the activities they are designed to engage in. But not all entities wear their activities on their sleeve.[10] Still, the point remains that more determinate kinds of entities have more determinate kinds of activities in which they can engage. Take for instance the following sequence of increasingly determinate kinds of

---

[10] I thank Carlos Zednik for emphasizing this point, and suggesting the metaphor of wearing an activity on the sleeve.

gases: gas, noble gas, helium. There are some things we can say about gases in general, like the fact that they obey the ideal gas law. Noble gases are largely inert, though different noble gases will potentially interact (under special conditions) with different molecules. Helium is a determinate kind of gas—characterized by a specific kind of atom with a specific atomic weight, etc. (though even it is not fully determinate, as it has a number of isotopes). Helium can engage in certain activities, like lifting balloons, that not just any noble gas can. Classifications of kinds of gases may cross-cut: helium is a gas that is both inert and lighter than air—which makes it ideal for balloons. Hydrogen is lighter than air but not inert, while Xenon is inert but not lighter than air.[11]

This analysis I hope provides some justification of the weakly realist account of kinds of entities and activities that I am advocating. Following the models-first strategy, entities and activities are classified into kinds in virtue of us having abstract models that capture multiple entities or activities under a kind; but those models can be used to represent these things in virtue of real similarities between features of the model and features of the targets. These similarities justify us in saying that there really are kinds of entities and activities.

4.4 Abstraction and Kinds: Some Lessons from Software Engineering

To conclude this chapter I want to draw some lessons about abstraction and kinds from what may seem like an unlikely place—software engineering. Software engineering is a field that depends heavily on the use of abstraction, and designers of computer languages have had to think carefully and formally about how to provide tools to manage these abstractions. In particular, the approach to abstractions in what are called object-oriented program languages provides an illuminating way about thinking about abstraction in modeling entities, activities, and mechanisms.

Computer scientist Robert Sebesta defines "an abstraction [to be] a view or representation of an entity that includes only the attributes of significance in a particular context" (1999, 412). The key points in this definition are that abstractions are features of representations, that abstractions involve the omissions of some features of the entities represented in favor of others, and that what these favored or significant features are is context dependent. In a book on software design and analysis,

---

[11] While I have explained this account of kinds in terms of relations between determinates and determinables, some readers may wonder whether these relations are better thought of as relations between a genus and a species. The determinate/determinable and genus/species relations have similar characteristics, though in the case of the genus/species relationship, the idea is that there is a specific defining feature that can be conjoined to the genus to get the species—for instance, a human is an animal that is rational, or a triangle is a polygon that is three sided. I have borrowed the determinate/determinable distinction because it seems more general, and I doubt all relations between more or less determinate activities or entities could be captured under the genus/species relations. But exactly how to understand the differences between these different relationships is a matter of ongoing debate (Rosen 2010; Sanford 2014).

Grady Booch echoes these themes, providing an overtly pragmatic definition of an abstraction:

An abstraction denotes the essential characteristics of an object that distinguish it from all other kinds of objects and thus provide crisply defined conceptual boundaries, relative to the perspective of a user.   (Booch 1994, 41)

The dependence of abstraction on perspective is reflected in the phenomenon-dependence of decompositions that figured centrally in our discussion of parts and boundaries within mechanisms. The delineation of parts within a mechanism is always relativized to the phenomenon for which the mechanism is responsible; the phenomenon provides the context that allows one to abstract away from those features of the parts and their activities/interactions that are not relevant to the production of the phenomenon.

Advances in abstraction techniques over the last forty years has been crucial for the design and construction of increasingly large and powerful software systems. Of particular interest for our discussion has been the development of object-oriented programming languages (OOPLs) and the paradigms of object-oriented analysis and design. OOPLs like C++ and Java have syntactic and semantic features that facilitate abstract representations of systems, and object-oriented analysis and programming techniques help engineers and developers to construct abstract representations of systems and processes. Object-oriented programming languages were originally developed as modeling languages, and they are frequently used to code simulations that count, in the sense described in Chapter 3, as mechanistic models. By considering how abstraction is used in these models, we can get a better sense of how the models-first strategy yields us abstract kinds.[12]

Software engineers often use the term "mechanism" to characterize certain software constructs. Booch defines a mechanism in the following way:

Mechanism: A structure whereby objects collaborate to provide some behavior that satisfies a requirement of the problem.   (Booch 1994)

This echoes, accidentally but strikingly, a well-known characterization of mechanisms from Bechtel and Abrahamsen.

A mechanism is a structure performing a function in virtue of its components parts, component operations, and their organization. The orchestrated functioning of the mechanism is responsible for one or more phenomena.   (Bechtel and Abrahamsen 2005)

There are, of course, differences between these definitions that reflect differences between the goals of the scientist and the software engineer: the software engineer is chiefly concerned with building mechanisms rather than representing or explaining

---

[12] I shall not elaborate on the point here, but I believe that a very large class of computer applications that we would not traditionally think of as models or simulations, including database applications and control systems, require the abstract representation of mechanisms.

them. But in the case of computer models or simulations this distinction collapses. The purpose of the mechanism constructed in software is to represent and explain the behavior of a mechanism in the natural or social world.

A number of features of object-oriented languages facilitate abstraction and thereby generate kinds. Foremost among these is the concept of an object itself. An object-oriented program consists of a set of declarations and instructions that describe various classes of objects. OOPLs make a distinction between object classes and object instances. A set of declarations and instructions will define a class, and a program may create many instances of the class. Members of an object class form a kind of object within the program.

A definition of an object class has two parts: an interface and an implementation. The interface defines the ways in which an object can interact with other objects. It specifies an object's "methods"—which we would, in the terminology of this book, call its activities. The methods in the interface characterize the capacities of the object, but they do not describe the mechanisms responsible for those capacities. These mechanisms are specified in the implementation. OOPLs enforce what is called encapsulation. This means that they prevent a program from interacting with an object or changing its properties, except via its interface. The interface thus creates the boundary of the object. This conception of object boundaries closely parallels the conception of entity boundaries that I sketched in Section 2.5. Object boundaries are not necessarily characterized by a spatial location, but by methods in the interface that afford opportunities for other objects to interact with it. Perhaps the most significant difference is that the encapsulation built into OOPLs allows programmers to build software mechanisms in which aggregations of objects are strictly rather than nearly decomposable. The ability to separate interface from implementation is one major tool for abstraction in OOPLs. It allows one to characterize what an object does without filling in the details of how it does it.

The other key feature that OOPLs provide to facilitate abstraction are the facilities for arranging objects into classes and subclasses. In writing an object-oriented program, one first writes code to define a class of object, and then one can, elsewhere in the code, create instances of the object. This is how an OOPL distinguishes between types and tokens. Object classes form hierarchies from the most abstract object class (the class *object* itself) down through more and more determinate subclasses. More concrete classes inherit and extend the methods of the more abstract classes. An object class hierarchy, once built, provides a hierarchy of object kinds.

OOPLs were originally designed for the purposes of building simulations of both natural and engineered systems. OOPLs are thus model-building languages. While OOPLs can be used to build models of a wide variety of sorts, they are particularly well suited to building models that simulate the behavior of a whole system by simulating the activities of and interactions between each of the system's parts. The use of such models, typically called individual-based models or agent-based models (IBMs or ABMs), has exploded in the last twenty years, and they have been used in a variety of natural and social scientific contexts, including modeling of the spread of infectious diseases, of flocking and schooling behavior in fish and birds, and of patterns of forest succession (Grimm et al. 2005; DeAngelis and Mooij 2005). IBMs are mechanistic models because these models show how a collection of individuals give rise to some phenomenon that is the behavior of the system as a whole.

To get a more concrete sense of the role of abstraction in IBMs, it would be useful to consider an example. Uri Wilensky's (1997) Wolf Sheep predation model is a highly idealized simulation of the dynamics of predator-prey relationships that has chiefly been used as a pedagogical tool in undergraduate classrooms (Wilensky and Reisman 2010). It is written in NetLogo, a language and modeling environment for agent-based simulations. Figure 4.1 contains a screenshot of the visual interface, which gives a quick idea of how the simulation works.

The simulation takes place in a world consisting of a grid. On this grid, three kinds of objects can be found: grass, wolves, and sheep. Things happen in the model in discrete time steps: grass grows; sheep eat grass; wolves eat sheep. Metabolism for wolves and sheep is measured by a parameter called energy. Wolves and sheep each use a certain amount of energy with each click of time. If they run out of energy, they die. Also, both wolves and sheep reproduce. When the simulation is run, wolves and sheep will move around the world producing qualitatively familiar population dynamics. If the sheep population grows too large the sheep will either start to starve and die off due to insufficient grass, or it will lead to a growth in the wolf population, which will lead in turn to a crash in the sheep population, and then another crash in the wolf population.

The wolf-sheep model requires (at least) the following classes of object:[13]

World
Grass
Wolves
Sheep

The world is represented by the map. It contains a number of instances of grass, wolves, and sheep, each with a location on a grid. There is only one instance of the object class *world*.

Considerable conceptual and programming efficiency can be achieved by organizing these object classes into hierarchies. Here is one way to do it:

Object
    World
    Organism
        Grass
        Animal
            Wolf
            Sheep

---

[13] I have diverged a bit from the way in which the actual model is coded and in the names I am giving to classes, but this description characterizes the essential features of the model implementation.

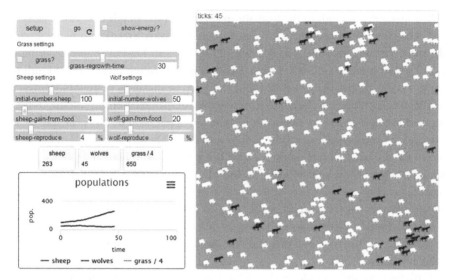

Figure 4.1 Wolf sheep predation model interface. Screenshot generated from the web-based version of the simulation at http://www.netlogoweb.org/.

Indentation reflects subclassing, so a wolf is an animal, which is an organism, which is an object. Note the difference between the subclass relation and the containment relation. Subclasses exhibit what software engineers call an "is-a" relationship to their parent classes. A wolf is an animal, and an animal is an organism. On the other hand, the world does not have a subclass, but it does contain other objects as parts. This is what software engineers call the "has-a" relation. The world has grass, has wolves, and has sheep. The is-a relation is the relationship we characterized above as the relation between determinates and determinables. To be a wolf is to be an animal, not *simpliciter*, but in a specific way. The has-a relation is a part-whole relation: in this case, the world is an object that contains parts—various instances of grass, wolves, and sheep.

Each class is defined by a set of declarations that describe the properties of the objects, and the activities that it can engage in. For instance, the abstract class *organism* has one important property—location—and two activities (methods) it can engage in—it can be born (come into being) and it can die (cease to be). In the wolf-sheep model grass is born a set time after it dies and lives until it is eaten. The class *animal* has all of the properties and methods of the class *organism*, but it adds three more: eat, move, and reproduce. The subclasses *wolf* and *sheep* will "override" these methods to specify how individuals in these subclasses go about these activities. For instance, eating for wolves is a matter of eating sheep, while eating for sheep is a matter of eating grass. The sheep will go hungry (and lose energy) if they are on bare ground, while the wolves will go hungry if there are no sheep around to eat. Reproducing and dying are handled similarly for wolves and sheep. In both cases you die if you run out of energy and reproduce with a certain probability within each time interval.

The hierarchy of kinds in the wolf-sheep simulation is a simple one. Every class of object has exactly one parent class. Classes ordered by the is-a relation form a tree with the most abstract class *object* as its root. In OOPLs, this kind of simple ordering is called single inheritance. But we have already observed in our discussion of kinds of entities and activities that concrete kinds can often be instances of multiple abstract, less determinate kinds. To facilitate representation of such relationships, some OOPLs support *multiple inheritance*. We can imagine an application of this principle to an extension of the wolf-sheep model: wolves and sheep are certainly kinds of animals, but so too are males and females. A particular female wolf is both an instance of the kind female and of the kind animal. To handle this kind of case in a simulation, OOPLs must allow concrete objects to inherit properties or behaviors from multiple parent classes. So a female wolf would be an instance of both the *female* class and the *wolf* class, and would inherit from these classes different methods (activities)—from the *female* class, for instance, birthing and nursing, and from the wolf class non-sexually differentiated wolf behaviors.[14]

With the wolf-sheep model in view, let us consider again how the models-first strategy yields entity and activity kinds, as well as mechanism kinds. According to that strategy, whether two mechanisms (or entities or activities) are of the same kind depends upon whether they can be represented by the same model, and whether any actual system can be represented using any particular model depends upon the stringency of the similarity requirement. The wolf-sheep model was not built with any particular animal populations in mind. Its purposes are primarily pedagogical, and it is at best a how (very)-roughly model of any particular ecosystem. Certainly it will not provide quantitatively useful predictions about population dynamics in any actual system. What it does do is illustrate some qualitative features of those dynamics that are discernable in many trophic systems. When a food source expands, populations of consumers will follow, leading to depletion of the resource and a contraction of the population. The model will also illustrate the importance of minimum densities. In the wolf-sheep model, for instance, wolves must find sheep, and if the sheep population gets small enough, the random search for sheep will come up empty and the wolf population will go extinct. The sheep population will then grow unchecked until they start to run out of grass; if the rate of grass growth is slow enough, this may lead to the extinction of the sheep. The wolf-sheep model illustrates the principle that false models can yield true theories (Wimsatt 2007, ch. 6), or, more precisely, qualitatively accurate mechanism-dependent generalizations of broad scope.

---

[14] This example oversimplifies things both in terms of the design of programming languages and in terms of the example. For one thing, not all OOPLs support multiple inheritance: Java, for instance, uses an alternative language feature (called interfaces) to get a similar effect. Also, it would in fact be very difficult to create an abstract class female, because there is very little that we can say abstractly and generally about the characteristics of the different sexes within the animal kingdom. Birthing and nursing is something that female wolves do, but those are species-specific (or at least specific to a lower level taxon like mammals).

While I have emphasized the importance of modeling goals in determining whether a model is a good enough fit to count as a model of an actual target, that determination also very much depends upon objective features of that target. Whether or not the abstractions and idealizations embodied in the model are permissible will depend upon whether the absent or misrepresented features are important difference-makers with respect to the phenomenon to be explained. For example, the wolf-sheep model assumes that reproduction is asexual, even though many of the predator and prey species to which this model might otherwise apply are sexually reproducing species. The supposition of this model (and of many predation models) is that this idealization is unproblematic because only the rate and not the manner of reproduction makes a difference to the phenomenon under consideration. If this supposition is true, the model justifies us in treating the target as an instance of a (weakly real) kind represented by the model.

4.5 Conclusion

The sciences are concerned with generality—kinds of things, and generalizations we can make about them. But mechanisms are heterogeneous particulars, complexes of entities and activities located in particular places and times. In this chapter, I have argued that models are what allow us to classify these particulars into kinds. Different particulars can be treated as of the same kind because they can be represented by the same model, or because they are treated as being of the same kind within a model that can be used to represent some larger phenomenon. So, for instance, we are justified in treating a pair of actual populations as instances of a wolf-sheep trophic system to the extent that the interactions between wolves and sheep in the model and the resultant system-level population dynamics resemble the interactions between predators and prey in the actual system, and their resultant system-level population dynamics.

This account is weakly realist. Kinds require models, and these must be constructed and deployed. How this is done depends upon scientists' goals, interests, as well as their cognitive capacities and technological resources. Nonetheless, nature places real constraints on which particulars can be represented by a model. Models can only be useful to the extent that there are objective similarities between the model and the target, and when a model can be used to represent two different targets, there must be similarities between those targets that license treating the targets as instances of a kind.

These objectively grounded similarities provide enough realism to do much of the work traditionally required of kinds. Entities, activities, and mechanisms of the same kind will behave in similar ways and respond similarly to interventions. These similarities ground the non-accidental but mechanism-dependent generalizations that, on the mechanist view, take the place of laws. In the next chapter, I will substantiate these claims by elaborating the classificatory schema introduced in Section 4.2, and showing how a variety of rather different kinds of mechanisms will fit within it.

5

Types of Mechanisms

In this book, I have argued for an expansive conception of what a mechanism is—one that is embodied in the definition of minimal mechanism. There is both a descriptive and a metaphysical rationale for this choice. Descriptively, my aim has been to capture the great diversity of things that scientists—both social and natural—call mechanisms. I still believe that there is something called general philosophy of science, and I think an understanding of mechanisms is essential for that project. The metaphysical rationale is that an expansive conception of mechanism will allow us to use a mechanistic framework to investigate some quite general questions in the metaphysics of science—most notably questions about the nature of causal and constitutive relationships between the entities, activities, and events one finds in the world.

As I noted at the beginning of Chapter 2, others have suggested that the metaphysical and descriptive projects should be disentangled (Andersen 2014a; Levy 2013). Most commonly it is thought that a narrower definition of mechanism is preferred for the descriptive project. Mechanistic science then would be a distinctive kind of science with its own special methods and explanatory forms. A prima facie benefit of a narrower conception of mechanisms is that it should allow researchers to characterize in richer detail the particular methods of mechanistic discovery, representation, and explanation. For instance, to the extent that mechanisms involve localizable functions of parts, decomposition and localization will be distinctive strategies for understanding how mechanisms work (Bechtel and Richardson 1993). A broader conception, it might be thought, would prevent one from making meaningful generalizations about such matters.

The difficulty with a narrow conception of mechanisms is that one must decide just how to do the narrowing. The most frequent suggestion (Baetu 2013; Andersen 2012) is that mechanisms must be regular. The idea is both that mechanistic sciences are in the business of explaining regularities in nature, and that in the absence of regularities it is impossible to find evidence for mechanisms. But, as these authors recognize (see also DesAutels 2011), there are quite a number of distinct senses in which a mechanism might be regular, and this regularity will come in varying degrees.

But even if one could solve the difficulties in drawing a clear line between regular and irregular mechanisms, it is reasonable to wonder why this feature or features is taken as the distinctive marker of "scientific" mechanisms. The kinds of mechanisms investigated by the sciences are heterogeneous across a variety of dimensions. I believe it is better to occupy ourselves exploring this space of mechanisms than trying to find a demarcation criterion (in terms of regularity or otherwise) to distinguish mechanistic from non-mechanistic science, or metaphysical from scientific conceptions of mechanisms. It is with the goal of exploring this space that I offer in this chapter a set of classifications for identifying different types of mechanisms.[1] As I have suggested in Section 4.2, the major dimensions can be read off the definition of minimal mechanism: one can distinguish different types of mechanisms by the kinds of phenomena they produce, the kinds of entities of which they are made, the kinds of activities in which those entities engage, the different ways in which entities and activities are organized to produce the phenomena for which the mechanism is responsible, and kinds of etiologies—the different ways that mechanisms can come to be.

Already in Chapter 4 I have discussed how entities and activities can be classified into kinds, so I will focus in this chapter on the phenomena, organization, and etiology. Within each of these dimensions I shall identify a number of features that can be used to mark different types of mechanisms. I choose this particular set of features because they can help identify similarities and differences between the kinds of mechanisms that operate within domains of scientific investigation, and, relatedly, because they illuminate topics that philosophers of science have found important. This collection of features is not meant to be exhaustive; indeed, I cannot imagine how an exhaustive collection could be constructed.

The features of mechanisms whereby they can be classified into kinds can be broadly classified into two sorts—what I will call material and structural features. Materially similar mechanisms will be similar because they are made of or make the same or similar stuff, while structurally similar mechanisms will be similar because they are structured or organized in the same way. As an example of material similarity, consider organic chemical reaction mechanisms. The essential feature of this kind of mechanism is that most of the entities involved in these will be organic molecules. If there is not carbon, there is not an organic chemical mechanism. And while I frame material similarity in terms of stuff, the interdependence of entities and activities entails that material similarities lead to similarities in activities. On the other hand, feedback mechanisms are a kind of mechanism defined by structural features. Negative feedback mechanisms are kinds of mechanisms that can be characterized by a certain organization of magnitudes—as one magnitude increases it drives an increase in a second magnitude, which feeds back to inhibit the first. Negative feedback is everywhere—in biochemistry, in electrical circuits, in economic markets, in social relationships among human beings.[2]

---

[1] I use the term "type" and "kind" interchangeably. The term "kind" is often more metaphysically freighted, but on the weakly realist view I have offered, there is no difference.

[2] The distinction between structural and material features is not absolute, and indeed what we think of as the material features of things often depends heavily upon their microstructure. For instance, to classify a molecule as organic is to say something about the material of which it is made. But a kind of matter like carbon is the kind of matter it is because of the structure it has. Moreover, material characteristics depend not just upon microstructure, but also context. It may be a material fact about oxygen that it is a gas, but a particular sample of oxygen's state (gaseous, liquid, or solid) depends upon the environment in which it is found.

Mechanisms are material things that are organized in such a way that they produce phenomena—so they will exhibit similarities and dissimilarities with respect both to material and to structural features. My hunch, which I will return to at the end of this chapter, is that this interplay between material and structural features accounts for certain features of the structure of scientific inquiry. Roughly, disciplines are often characterized by the material mechanisms they study, but different disciplines may appeal to common tools to understand structural similarities in materially distinct mechanisms.

5.1　Types of Phenomena

We observed in Chapter 2 that the term "phenomenon" as used in the characterization of minimal mechanism is meant to encompass things in a variety of ontological categories for which mechanisms may be responsible. There are mechanisms that underlie activities or interactions (like the mechanism by which the heart beats or the mechanism by which one animal signals another), mechanisms responsible for events (like the mechanism responsible for cell division), mechanisms that produce objects or stuff (like mechanisms that synthesize proteins, or refine gasoline), or mechanisms that sustain states (like mechanisms that regulate temperature in warm-blooded animals). We may also speak of mechanisms responsible for capacities or dispositions to act or interact.

But apart from these broad categories, we can classify mechanisms by the more specific sorts of activities, interactions, events, objects, stuff, or states for which they are responsible. This may be our most common way of classifying mechanisms. Certainly this is the case for machines we build: bread makers are mechanisms that make bread; electrical generators are mechanisms that produce electricity; vacuum cleaners are mechanisms that suck things off the floor. We also classify naturally occurring systems and processes in this way. We speak of mechanisms for protein synthesis, which are diverse in their structures, operations, and organization, but which share the common feature of producing proteins as their end-products, where proteins are molecules defined by the kind of chemical components of which they are made. Similarly, what makes an animal a carnivore or herbivore has to do with what kinds of materials its digestive mechanisms take as inputs. Or consider psychological mechanisms like the fight-or-flight reflex. We identify the operations of this kind of mechanism by the fact that a certain kind of stimulus (seeing a predator) produces a certain kind of response.

This form of classification is important because it allows us to understand how different mechanisms may be responsible for the same kind of phenomenon. From the mechanistic viewpoint, when we say that some phenomenon is multiply realizable, we are saying that a phenomenon of that type can be produced by different types of mechanisms, or that different types of mechanisms underlie different instances of the phenomenon. Similarly, when we are considering mechanistic processes, identifying products of these processes by their type allows us to make sense of the fact that the same type of product can be produced by different types of processes. And if (as I shall elaborate in the next two chapters) we understand these mechanistic processes as underlying causation, it makes sense of the idea that one kind of effect can have many kinds of causes.

Of course this kind of classification of mechanisms only works to the extent that we can successfully classify phenomena to be of the same kind. I have given a basic account of how to do this in Chapters 3 and 4: We may classify the phenomena mechanisms produce or underlie (entities, activities, etc.) in more or less determinate ways. This is a matter of giving more or less abstract models of the kind. Whether the two different kinds of mechanisms can cause or realize the same kind of phenomenon depends crucially upon how fine-grained the descriptions of the phenomena and mechanisms are.

Much more could be could be said here, but the main point I wish to make is that as a matter of fact, mechanisms are often classified by what they make or do. This is an example of what I am calling classification based upon material similarity—mechanisms for synthesizing proteins or for brewing coffee. For the remainder of this section, I want to consider more abstract and structural forms of classification of phenomena.

5.1.1 Constitutive and Non-Constitutive Phenomena

When a mechanism produces a phenomenon, sometimes the phenomenon can be characterized in terms of the activities of or relations between one or a few parts of the mechanism, while in other cases the phenomenon can only be characterized in terms of activities of the mechanism as a whole. I shall call phenomena of the first kind "non-constitutive phenomena" and phenomena of the second kind "constitutive phenomena."

As an example of a non-constitutive phenomenon, recall the bicycle drivetrain from Chapter 3. The drivetrain's primary behavior is non-constitutive because that behavior can be described by showing how changing the state or behavior of a few parts changes the behavior of another part. Specifically, there is a function from the rotation rate of the crank and position of the gear levers to the rotation rate of the hub. The drivetrain is a mechanism for making the hub (which is just one part of the system) turn at a certain rate. Contrast this kind of mechanism with the mechanism by which a muscle contracts. Here it is the muscle as a whole that contracts, not just one part of the muscle, so the muscle's behavior is constituted rather than caused by the activities and inter-actions of its parts. The distinction between constitutive and non-constitutive phenomena is really about different kinds of relations that obtain between a mechanism and its phenomenon, and it is the distinction that Craver and Darden call out when they distinguish between mechanisms that produce phenomena (non-constitutive) and mechanisms that underlie phenomena (constitutive).

While non-constitutive phenomena can be characterized in terms of the behavior of one or a few parts, non-constitutive phenomena are still produced by the whole mechanism. For instance, in the case of the bicycle, the non-constitutive phenomenon—characterized in terms of the functional relations between the rotation rates of the pedals and the hub, along with the position of the gear levers—is nonetheless a behavior of the system as a whole. Without all the parts of the mechanism, like the frame, chain, derailleurs, and sprockets, the non-constitutive phenomenon could not be produced.

While the examples of constitutive phenomena we have so far discussed are the products of mechanistic systems, phenomena produced by one-off causal processes can also be constitutive. Suppose for example that I am hiking through the Vienna Woods and I kick a pebble down the hill. Somehow this pebble manages to strike objects on its way down in such a way that it produces a rock slide. The entities involved in this one-off process are the rocks, but the phenomenon produced is the rock slide, which cannot be attributed to any rock, but is a constitutive phenomenon.[3]

5.1.2 Recurrent Phenomena

Within the recent philosophical literature on mechanisms, the phenomena that are typically discussed are recurrent phenomena. Whether the phenomenon be chemical, biological, or social, it happens a lot, and occurs always or for the most part in the same way. But minimal mechanism allows for mechanisms to be responsible for phenomena that do not recur. So one useful way to classify mechanism-dependent phenomena is by whether and in what sense they recur. A number of cases should be distinguished.

First, there are *one-off phenomena*. A one-off phenomenon is the product of a mechanistic process that occurs just once. For instance, in the spring of 2013 a combination of snow melt and heavy precipitation in regions surrounding the Danube river led to the water reaching a high water mark on the night of June 5, reaching a flood level of 8.09 meters measured at the town of Korneuburg, located just upriver of Vienna. The entities involved here—weather systems, tributaries, rivers, etc.—produced this single event, and the particular timing and level of this flood at this place was highly dependent upon the features of these entities, activities, and interactions.

Second, there are *recurrent phenomena*. Often it is the case that a particular system repeatedly engages in the same kind of process, producing the same kind of phenomenon. The Danube, for instance, continually moves water from its sources to its mouth, and its water levels repeatedly rise and fall, and reach flood stages whenever volumes of overall precipitation and melt in the river's watershed exceed certain rates. Because

---

[3] While a constitutive phenomenon is a phenomenon that can only be attributed to the mechanism as a whole, to say that the phenomenon is constitutive is not to say anything about how the parts are organized so as to be responsible for the phenomenon as a whole. Constitutive phenomena may be nothing more than a "sum of the parts" behavior, or they may depend on more complex forms of organization. To use Bill Wimsatt's term, constitutive phenomena may be either aggregative or non-aggregative (emergent) (Wimsatt 2000).

this kind of phenomenon recurs in the very same system (e.g., the Danube watershed), we can refer to a phenomenon of this kind as a *same-system recurrent* (or repeatable) *phenomenon*.

In addition to recurring within systems, a phenomenon may recur across different systems of similar type. Continuing with our example, flooding is a phenomenon that recurs repeatedly in many different river systems. It is not just the Danube that floods, but the Nile and the Mississippi and many other rivers. In these cases, there will be common kinds of variables that determine how rainfall and other factors affect the water level and flow rate through these systems. When a phenomenon exhibits recurrence across multiple systems, we can call this a *cross-system recurrent phenomenon*.

Same- and cross-system recurrence are distinct and we may see kinds of phenomena that exhibit one or both varieties. Some quick examples: Red blood cells are systems that exhibit recurrent phenomena in both senses. Red blood cells have a lifespan of around 100 days, during which time they circulate continually in the bloodstream, absorbing oxygen within the lungs and releasing it in less oxygen-rich regions of the body. At the same time a single human being will have trillions of red blood cells, as will every other human being and every member of a wide range of other species. Red blood cells exhibit both same- and cross-system recurrent behavior. On the other hand, some systems are unique and non-recurrent, but exhibit repeatable behavior. The United States Postal Service is one of a kind, but it exhibits (or at least aspires to exhibit) a robustly and frequently repeatable phenomenon of mail delivery. As their motto goes, "neither snow nor rain nor heat nor gloom of night stays these couriers from the swift completion of their appointed rounds." Finally, we can find many examples of systems that exhibit a behavior just once (if at all), but of which there are many copies. A single oak tree will produce many acorns, but an acorn will develop into an oak at most once, if at all.

A category of phenomena related to recurrence are *mechanism-dependent dispositions*. Mechanistic systems have dispositions that only manifest themselves under suitable triggering conditions. The car will only start if you turn the key, and the bee will only sting if threatened. The manifestation of a mechanism-dependent disposition may occur rarely, frequently, just once, or not at all. The car starts each time you turn the key (always or for the most part), while the bee stings at most once, as the bee's stinger is planted in its target and the bee dies. And if there are many tokens of a type of mechanistic system, like the bee, then the manifestation of a type of phenomenon can be one-off within a system, but recurrent across systems. A single bee stings at most once, but many bees sting.

A mechanism's phenomenon can be categorized as one-off or recurrent only with respect to a representation of the phenomenon. Since the very same concrete mechanistic process or system can generally be represented by different models, there is no one type of mechanism that it is. For instance, whether one treats the Danube flood as a one-off event or as an instance of a recurrent process depends largely upon descriptive grain. Relative to a fine-grained description, one that characterizes precise water levels and flow rates in various locations over a series of days, the flood in June 2013 will certainly never be repeated. Nonetheless, one could construct a generalized model of Danube flooding that estimates crest levels and flow rates at various points on the river as a consequence of various inputs (setup conditions). Models of this sort allow officials to predict with reasonable accuracy the time and level of water crests. Other models may be yet more general. For instance, engineers seeking to understand the consequences of changes to land use might construct generalized models for understanding the consequences of building dams or levees on the ways in which a river system will respond to high levels of precipitation. Such models can be created with parameters that would allow them to represent water flow in a variety of river systems under varying conditions. Different models are created for different purposes—for instance, in this case to predict as precisely as possible water crests on a particular river, or to guide land use policy within the watershed. The models-first strategy implies that pragmatic considerations will set descriptive grains and idealizing assumptions, and these features will determine whether a token system or process will be described as recurrent or one-off, or as exhibiting stable dispositions. This fact does not mean that we cannot find objective differences between some processes and others. It will, for instance, be a matter of empirical fact how well a generalized model of water flow in the Danube will be able to capture the dynamics of the watershed over multiple years.

To say of some mechanism-dependent phenomenon that it is recurrent is not to say anything about how regular that recurrence is. Mechanistic systems with stable dispositions may be triggered only irregularly, and systems that exhibit recurrent behavior need not behave the same way whenever triggered.

5.1.3 The Functional Structure of Phenomena

As I suggested at the outset of this chapter, to say that two mechanisms produce, underlie, or maintain the same kind of phenomena could, in its broadest sense, mean one of two things. On the one hand, one could be saying that the mechanisms were responsible for producing, underlying, or maintaining the same kind of stuff (or states of the stuff or changes to the stuff). They might both produce proteins, or reduce joint inflammation, or maintain a constant temperature within your home. On the other hand, one could mean something more abstract. There may be similarities in the structure of the phenomena that are independent of what particular kind of stuff the mechanisms are producing, underlying, or maintaining. When we say of a mechanism's phenomenon that it is recurrent, or regular, or periodic, or chaotic, we are saying nothing about *what* it makes or does, but are saying something about the patterns in those makings or doings; that is what I mean by characterizing that phenomenon structurally.

The nature of structure is notoriously elusive, and the varieties of structure are literally endless. So I will not offer any theory of structure or offer a systematic account of kinds of structure. I will content myself with the more modest task of showing how kinds of structure can be found in models of the phenomena for which mechanisms are responsible. In Chapter 3 I suggested that we can think of a model of a mechanism as having two parts—a phenomenal description, and a mechanism description; the phenomenal description is an account of what the mechanism does, while the mechanism description is an account of how it does it. Structurally similar kinds of phenomena will have structural similarities in their phenomenal descriptions. Models come in many forms—material models, mathematical equations, diagrams, computer simulations, etc., but to talk about structural similarities among mechanism-dependent phenomena, it will be easiest to focus on mathematical models. In these models, the phenomenal description can typically be given by some function, and one can classify materially disparate phenomena into kinds if these functions have common structural features. This is what I mean by the functional structure of phenomena.

We begin with *input-output mechanisms*. Input-output mechanisms are systems whose behavior can be described by a functional relation between input and output variables, e.g.:

$$O = F(I_1, \ldots, I_n).$$

In input-output mechanisms, inputs *produce* outputs. They exemplify the relationship implicit in Machamer et al's (2000) characterization of mechanisms as moving from start-up to termination conditions. Mechanistic systems with stable dispositions are examples of input-output mechanisms, in the sense that the triggering conditions for the manifestation of the disposition are inputs, while the manifested behavior is the output. The honey bee, when appropriately triggered, will sting. Similarly, the bicycle drivetrain is an input-output mechanism relating the input rotations of the crank to the output rotations of the hub.

The process by which a system produces an output in response to inputs will involve the activities and interactions of its component parts; these will take time, but the temporal dimension will only sometimes be represented in the description of the phenomenon. For some phenomena, the input-output relation is adequately captured by a simple mapping from input to outputs. I can, for instance, provide a phenomenal description of a Coke machine that maps various combinations of inputs to outputs. That mapping says things like "if you insert a dollar and push the button labelled 'Coke' a Coke will come out of the slot." For other phenomena, the temporal dynamics of the system may be of much more interest. In representing the action potential, for instance, scientists do not simply describe its setup and termination conditions (e.g., a threshold stimulus and resting potential), but instead characterize the phenomena in terms of change in membrane potential over time. In cases such as this, we might represent the input-output functions using differential equations with time derivatives, or by adding time as another variable to the input-output function, e.g., $O = F(I_1, \ldots, I_n, t)$.

In accordance with the models-first strategy, whether or not a mechanism is "dynamic" is in the first instance a feature of a model of that mechanism, and not of the mechanism itself. Whether temporal features of the phenomena will be included in the model will depend crucially upon the purposes of the modeler. That being said, the modeler's purposes are inevitably shaped by features of the mechanism and its relation to its context. As an example, consider a very familiar input-output mechanism—the computer at which I am now typing. One of the phenomena produced by the computer is (in appropriate contexts, like when a word-processing window is open) a mapping from presses on a keyboard to the display of characters. When I press the "s" button I see an "s." While this process takes time, for most purposes, all that is needed to describe this phenomenon is the input-output mapping, treating the operation of the mechanism as instantaneous. The reason is that in current computers the time lag between typing and displaying is on a timescale that is imperceptible to human users. Contrast this with downloading files off the internet. Here, there are perceptible time delays and variations, and engineers will therefore have interests in describing the performance in terms of its temporal characteristics. One way of describing how onto-logical constraints shape the interests of those representing mechanisms with models is to say that mechanisms will typically be embedded in larger mechanistic contexts, and that those contexts will determine what features of the behavior of the embedded mechanism are causally relevant. Computers are input-output mechanisms that are embedded in larger mechanisms involving interactions between computers and their users. The features of the behavior of computers that will be of interest to a modeler will be just those features that make a difference to the way that computers and users interact. We have the same situation with the action potential. The reason scientists are interested in the temporal characteristics of the action potential is that timing makes a difference in the operation of the nervous system.

In contrast to input-output mechanisms, some mechanisms are responsible for producing a pattern of behavior in the absence of input from outside the system. We might call these *no-input/output* mechanisms. Obvious examples of such mechanisms are systems that exhibit oscillatory or other forms of periodic behavior. Schematically we can represent them using equations of the form:

$$O = F(t).$$

Such systems are idealized as closed systems whose dynamics follow solely from the activities and interactions of its parts. Here are a few examples: oscillating springs, pulsars, watches, circadian rhythms, and predator-prey systems. Let me briefly elabor-ate on features of a couple of such systems to examine their features. First, consider mechanisms for maintenance of circadian rhythms.[4] Circadian rhythms are, broadly speaking, patterns in the physiology and behavior of organisms that are organized in twenty-four-hour cycles. There are many such patterns. For instance mammals exhibit circadian oscillations with respect to body temperature, blood pressure, digestion, sleep, and alertness. The crucial feature of circadian cycles that allows for the treatment

---

[4] Bechtel and several of his collaborators made an extensive study of how the mechanisms underlying various circadian processes are explored and modeled, and my brief description is drawn from their work (Bechtel 2009; 2010; 2012; Bechtel and Abrahamsen 2010; Kaplan and Bechtel 2011).

of them as endogenous oscillators is that they maintain the rhythm in the absence of external cues. They are not immediate responses to daily changes in the environment like periods of light or darkness or changes in temperature. It is important again to emphasize when we characterize phenomena like circadian rhythms as operating independent of environmental inputs, we are saying something about the models we can use to represent such systems, and not saying that they are actually closed systems. In fact, it is well known both that circadian cycles are maintained to some degree independently of the environment, but also that cycles can be entrained by environmental cues (so-called "Zeitgebers"). To concentrate on a familiar case, we get jet lag because our circadian cycle is internally maintained, but we get over jet lag because in time environmental cues lead to adjustments in the clock. Different scientific investigations will focus on different aspects of the cycles—some on the endogenous clock and others on the entrainment mechanisms.

A second example of an oscillatory mechanism comes from ecology, in the form of oscillations in the size of predator and prey populations. Lotke-Volterra models of predation characterize population changes in terms of lagging oscillations between predator and prey cycles. The model is a phenomenal one, but it is suggestive of the basic features of a mechanism: growth in prey population increases the food supply for predators, leading to growth in the predator population. As that population grows, the population of prey will fall due to increased predation, leading to scarcity of food for predators, and with it a lagging decline in the predator population. The pattern of oscillation is again a feature of the model that is at best an idealized description of the dynamics of actual populations. Actual population fluctuations will not match the curves of the Lotke-Volterra models, and population levels will be driven by a host of factors not included in the model. Still, to the extent that the Lotke-Volterra models provide a useful if idealized characterization of the phenomena, we are justified in calling the mechanism an oscillatory one.[5]

In addition to kinds of mechanisms that produce phenomena from inputs, and those that exhibit dynamics independent of inputs, there is a third class of mechanisms in which the phenomenon in question is the maintenance of a stable state in light of changes to the mechanism's inputs or changes in its environment. These are *homeostatic mechanisms*, what Craver and Darden (2013) call mechanisms for maintaining phenomena. We also might for the sake of parallelism call these "input/no-output" mechanisms, but this is misleading, both because homeostatic mechanisms are active and also because homeostatic mechanisms will dampen fluctuations but not eliminate them entirely. Homeostatic phenomena are ubiquitous in biological systems—for instance in mechanisms that maintain near constant body temperature in warm-blooded animals or mechanisms that maintain blood sugar concentrations within stable tolerances.

---

[5] Huffaker (1958) is a classic study that recreates the qualitative characteristics of predator-prey oscillations with actual organisms in a controlled experimental environment.

This threefold division between mechanisms that produce outputs as a result of inputs, mechanisms that produce outputs independently of inputs, and mechanisms that maintain stable outputs in light of varying inputs provides one useful way to characterize kinds of mechanistic phenomena, but it only scratches the surface of the functional structure of phenomena. More generally, if we can characterize some phenomena using mathematical representations that highlight the structure of that phenomena, then we can classify mechanistic phenomena into kinds based upon aspects of that structure. Here are just a few examples:

- We may characterize it in terms of the kinds of values variables describing the phenomena may take—e.g., logical (Boolean), discrete, or continuous.
- We may characterize patterns of dependence in terms of the functional form of relations between variables—e.g., linear, quadratic, exponential, etc.
- When the behavior of a system can be characterized by differential equations, we can classify phenomena by the character of their state spaces—e.g., by identifying phenomena with similar oscillations or equilibrium points.

All of these classifications refer to patterns in the models of phenomena. The kinds of variables used are features of a model, not of a concrete phenomenon itself. Also, they classify mechanisms on the basis of patterns in the phenomenon, without reference to the mechanism that produces it. A pure model of the phenomenon is simply a description of the functional dependencies of inputs and outputs or of the time evolution of the system's behavior. It is a description of what a black box does, not of how it works.

Stochasticity is a feature of some phenomena that falls under the general category of the functional structure of phenomena. To say that a mechanism's phenomenon is stochastic is to say that a model of the mechanism's phenomenon can be expressed in terms of probabilities or relations between random variables. For instance, the behavior of an input-output mechanism might be characterized as a conditional probability of an output given a particular set of inputs. Just how to interpret such probabilities is a complicated matter, about which I shall say more in Section 5.4.

5.1.4 Normal, Unusual, and Abnormal Phenomena

A last distinction in kinds of phenomena concerns phenomena that can be characterized as normal, unusual, or abnormal. The minimal conception of mechanism makes no such distinction among phenomena. One can pick any activity or event whatsoever and ask what mechanisms brought it about. What mechanisms are responsible for the flooding in my basement, for instance? Here there is no requirement that my basement usually floods or is supposed to flood. It just floods, and I want to understand the entities, activities, and interactions that led to the flooding.

The word "normal" has, in this context, two different senses. In the statistical sense, normal phenomena are usual or typical, and are contrasted with phenomena that are rare or unusual. In the normative sense, normal phenomena are by some standard right or proper, and are contrasted with phenomena that are abnormal, pathological, or somehow wrong. These senses are often intertwined. For example, it is normal in the statistical sense for children to enter the two-word phase of speech development between eighteen and twenty-four months, though some do so earlier or later. If a child enters this stage early, it is unusual but no cause for concern. If, however, they are delayed in entering this phase until much later, parents might worry that the child's behavior is not normal in a different sense. Here in the background is a conception of the proper rate of language development, and a concern that there is some breakdown of normal developmental mechanisms and the attendant concern that the child will not develop normal (in a normative sense) capacities for speech and communication. So the child's language development in this case is both unusual and abnormal.

The normative sense of normal phenomena is closely connected with what Garson (2013) has called the functional sense of mechanism. If something is a mechanism in the functional sense, it has a function—the way it is supposed to behave or the kind of phenomena it is supposed to produce—and with this, there is the possibility of malfunction. In the case discussed above, there is a normal course of language development (or, more properly, a normal range of variation within the course of language development) and when language delays get too far off this course, we would characterize these developmental mechanisms as malfunctioning.

Garson is certainly right that the functional sense of mechanism (and with it a normative sense of function) is "ubiquitous and useful." Functional ascriptions are very common in biology and medicine (and for that matter in psychology and other social sciences), and the distinction between normal function and malfunction can be useful, because it helps organize scientific inquiry and techniques for intervention. For instance, it can be helpful in medicine to first identify the mechanisms responsible for the normal function of some biological system, and then identify the various ways in which pathogens or environmental changes can interfere with or break the mechanism.

What I want to add here is that ideas of normal function and malfunction can also be useful in cases where it is not appropriate to speak of normative functions. Consider, for example, the following description of a recent outbreak of unusual behavior in oscillations of the sun's magnetic field:

The Sun's magnetic field, which varies primarily on a timescale of 11yrs, originates in the solar interior and so can be studied using helioseismology. The recent solar minimum between cycles 23 and 24 was deeper and longer than anyone expected. In a recent paper, Basu et al. use helioseismology to show that it was not just the minimum between cycles 23 and 24 that was unusual. In fact, the Sun was behaving strangely for the majority of cycle 23 and so, in hindsight, the recent unusual solar minimum could have been predicted.[6]

---

[6] http://www.uksolphys.org/uksp-nugget/30-could-the-unusual-solar-minimum-have-been-predicted/.

The phenomenon in question here is the oscillation in strength of the sun's magnetic field. There is a normal pattern of oscillation, and when the period of oscillation lengthens, it is said to behave unusually. The normal phenomenon is described by the model, and when the actual phenomenon varies from what is predicted by the model, scientists look to changes within the mechanism responsible for the phenomenon—here changes in helioseismic waves. More generally, when a system (or class of systems) exhibits recurrent behavior, it is useful to both characterize the normal ("always or for the most part") behavior, and then to look for changes in conditions that produce deviations from this normal behavior.

Much more could be said about the sources of normativity required to make the distinction between properly functioning and malfunctioning mechanisms.[7] For my purposes here, it is enough to observe that some mechanisms can malfunction while others cannot, and for some mechanisms it can be useful to distinguish between regular and unusual behavior, even if the unusualness does not imply any sense of malfunction.

5.2 Types of Organization

In addition to classifying mechanisms by what they do (their phenomena), we may classify them by how they bring about their phenomena. In characterizing the type of mechanism responsible for some phenomenon, we can appeal either to what the mechanism is made of—its constituent entities and activities (see Section 4.3)—or to how those constituents are organized. In this section, then, we take up a number of features of mechanistic organization that are useful in classifying mechanisms.

To describe a mechanism's organization is to characterize relations between parts and other parts, relations between parts and wholes, and relations between earlier and later stages. The relations can be spatial, temporal, and causal. At its finest grain, two mechanisms would have the same organization if there were a structure-preserving isomorphism between their parts, and isomorphisms between each mechanism and an ideal model. But there are no ideal models or perfect isomorphisms. Instead, we classify two mechanisms as having the same type of organization to the extent that there is at least a partial and idealized model that captures similarities between the organizations of the particular mechanisms.

The following subsections highlight several of the most significant organizational features by which we may fruitfully classify mechanisms. The features are fruitful in large part because the methods we use to discover, represent, explain, and control mechanisms will differ substantially depending upon which of these features they have.

Before turning to these features, I should observe that structural features of phenomena discussed in the previous section also count as organizational features of

---

[7] See Mossio, Saborido, and Moreno (2009) and Garson (2013) for recent discussions and references to standard accounts.

mechanisms. The basic distinction between mechanistic systems and mechanistic processes is an organizational one. Mechanistic processes are characterized primarily in terms of temporal stages, while systems are characterized in terms of componential as well as causal relations. Additionally, the distinction between constitutive and non-constitutive mechanisms describes an organizational relationship between the entities involved in the exhibition of the phenomena and the entities that are involved in the production of the phenomena.

5.2.1 Few Parted versus Many Parted

Some mechanisms have a few parts; some have a great number, and much about how we represent a mechanism in a model will depend upon how many parts it has. Technological artifacts typically have relatively small numbers of parts—small enough so it is practical to have a parts list. The bicycle drivetrain is an example of a few-parted mechanism. When the absolute number of parts in a system gets large, good engineering practice is to create more complex machines from relatively few modules, which themselves are made of relatively few parts.

Few-parted mechanisms can be found in natural and social systems as well. For instance, the mechanisms responsible for particular motions in humans or other animals will involve a relatively small number of bones, joints, tendons, and muscles. A social organization like a family unit or a gang will derive its capacities from the interactions between small numbers of actors.

In few-parted mechanisms, it is possible to represent each part of the mechanism, and to try to identify the activities and interactive relationships between each part. A hand surgeon will presumably need to know about the role of each bone, muscle, and tendon in the hand if she is to be able to understand functioning and malfunctioning hands, and when necessary to introduce repairs.

As the number of parts in a mechanism grows it becomes increasingly difficult to represent those parts individually and explicitly in a model. We see this, for instance, as we move to molecular biology. Take, for instance, the example of the action potential. Scientists understand that a crucial part of this mechanism is the voltage-gated potassium and sodium ion channels—but no one counts how many of these parts there are, and quantitative models of the behavior of the phenomena do not make reference to individual channels, but instead move to measures of the aggregate capacities of large collections of these channels. The Hodgkin-Huxley model, for instance, has terms that refer to the total ion current across a kind of channel. And though a mechanistic model of the action potential will acknowledge that this current passes through discrete gated ion channels, any representation of the overall action of the neuron will only characterize the operation of these channels in the aggregate.

There is no absolute distinction between few-parted and many-parted mechanisms, and what we count as a few will say a lot about our own cognitive, instrumental, and representational limitations. This is not a distinction that matters for Laplacean demons, but it matters greatly for the practice of science.

5.2.2 Uniform versus Functionally Differentiated Parts

In addition to the absolute number of parts in a mechanism, we can distinguish mechanisms that have many *kinds* of parts from those that have only a few. At one extreme, we can have a mechanism in which all parts are of essentially the same kind, with the activities and interactions they engage in being accordingly of the same kind. Contrast an hour glass and a wind-up watch. The grains of sand in the hour glass are uniform with respect to the characteristics relevant in their contribution to the hour glass's behavior. The watch on the other hand will have different parts with different functions and different kinds of interactions—some electrical, some chemical (within the battery), and some mechanical.

At issue here is not any absolute similarity of parts, but similarity with respect to capacities that give rise to the activities and interactions productive of the phenomenon. To see this consider an example of the mechanistic explanation of movement of bird flocks or fish schools. Realistic simulations of flocking behavior have been produced using agent-based models, which suppose that flocks consist of an ensemble of birds using identical rules for updating their flight path based upon the behavior of their neighbors (Bajec and Heppner 2009). If these models have correctly captured the kinds of interaction that are productive of the phenomenon in real flocks, then it shows that flock mechanisms are uniformly parted with respect to this behavior. With respect to other behaviors (mating, nesting, feeding, etc.), these very same birds may exhibit functional differentiation.[8]

5.2.3 Affinitive versus Induced Organization

What determines which parts of a mechanism interact with which others? Why does part A interact with part B, but not part C? There are broadly speaking two ways a mechanism may be put together that lead to two different kinds of organization, which I shall refer to as affinitive and induced. Consider two ways to organize a dinner party: one way is to assign seats; another way is to set out tables and just let people find their own seats. In the former, the organization is induced. Who talks to whom is determined by the host. In the latter, the organization will be determined in part by accident (where there is an open seat) but also largely based upon the various affinities of the people at the party—friends, family, colleagues, and so on will clump, or there may be other clumpings—the foodies, the football fans, the children, etc.

---

[8] Levy and Bechtel (2013) suggest that organized systems must exhibit two features: (a) different components of the system make different contributions to the behavior, and (b) the components' differential contributions are integrated, exhibiting specific interdependencies (i.e., each component interacts in particular ways with a particular subset of other components). In my terminology these are different kinds of organization. Feature (a) seems to be getting at what I here refer to as functional differentiation. Feature (b) is what I describe below as topological organization. Whereas Levy and Bechtel are using these features to distinguish organized from disorganized systems, I find it preferable to see these as different kinds of organization.

Paradigms of affinitive relationships are those between chemical reactants.[9] Within a solution, what molecules will react with what is not determined by where the molecules are located in the solution. There is, like in the party, enough mixing that the molecules will run into each other. But which molecules do react and create new products will depend upon their structures, and the affinities given to them by those structures.

There is a close connection between the idea of affinitive organization and self-organization (see Section 5.3.4), with the former typically being a kind of precondition of the latter. As I am understanding it, affinities are the capacities individual parts have to "seek out" and engage in interactions with other parts. Such affinities do not by themselves guarantee the emergence of larger organizational patterns, but they can make such patterns possible.

All mechanistic organization depends to some degree upon the existence of affinities—as parts must have the capacities to interact with other parts. Clocks are a paradigm of an induced organization, with each piece carefully placed in contact with just those pieces that it must interact with to produce the phenomenon. But still the gears must be of the right shape so that they can mesh and move each other. If you put me in the middle of a set of food stalls, I may be inclined by the scents to move toward one food stall rather than another—but if the stall only releases inert gasses, the vendor could put it right under my nose, and it would not move me.

One way to think of the degree to which the organization of a system is affinitive is the degree to which its ultimate organization depends upon its initial state (cf. Kuhlmann 2011). With strong enough affinitive relations, final organization will be insensitive to variations in setup. Some people will end up clumped together at a party, regardless of where they started. Mechanisms with these kind of affinities between parts will tend to be robustly self-organizing.

5.2.4 Topological and Functional Organization

Another way to characterize the organization of a mechanism is in terms of its topology, or the abstract patterns of connection between its parts. The natural mathematical tools for characterizing this organization come from graph theory. Graphs consist of nodes, which can represent parts or properties of parts, and edges (or links), which can represent causal interactions between the parts and their properties.[10] There are many kinds of graphs with many kinds of representational uses. I will not attempt to survey them (and indeed I do not think a systematic survey is possible), but only offer a few

---

[9] In fact, I draw the term from chemistry, where the idea of chemical affinity has a long history. As Baetu (2018) notes, the primary source of organization of molecular mechanisms is specific binding, which is determined by chemical affinities. Neither is it a new idea that organization of other kinds of entities can be affinitive. Goethe's 1809 novel *Elective Affinities* is motivated by the analogy between chemical and human affinities.

[10] I am eliding topology and graph theory, which are distinct mathematical fields. Both have to do with patterns of connectedness, though topology is more connected with geometrical relations on continuous surfaces. I concur with Huneman (2010) (who also elides these terms) that with respect to their use in modeling and scientific explanation, we can treat topology and graph theory as of a piece.

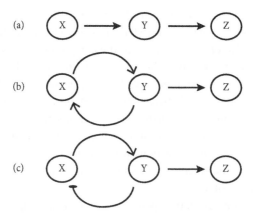

Figure 5.1 Three topologies: no feedback, positive feedback, negative feedback.

examples of topological organization. Consider first three topologies involving three nodes (Figure 5.1).

These topologies represent three basic sorts of network structures one might find in a mechanism with three parts. The first involves a linear acyclic graph, where X changes Y, which changes Z. The second involves positive feedback, so if some property of X is represented by a quantitative variable, increases in X will increase Y, which will feedback to increase X further. In the third graph there is a negative feedback loop, with X increasing Y, which in turn limits X. Note that these relationships could characterize relationships between parts of cars, or cells, or social organizations. Characterizing a mechanism in terms of a model of its topological structure allows us to see relationships between kinds of mechanisms that are materially very different and the structural dependencies allow us to explain the behavior of these mechanisms in a way that is independent of the concrete entities and their activities.[11]

Mechanisms can also be characterized by the relative density or sparseness of connection between nodes, and the degree to which nodes are clustered into neighborhoods. Levy and Bechtel offer as an example work by Watts and Strogatz (1998) on small-world networks. Their diagram representing the relation between regular lattices, small-world networks, and random networks is given in Figure 5.2.

Levy and Bechtel summarize the significance of their work as follows:

Watts and Strogatz... [demonstrated] that adding a few long-distance connections to a regular lattice... generates a network structure with both high clustering and short average path lengths between nodes. They termed such networks small-world networks, argued for their virtues in activities involving information processing, and determined that many real-world networks,

[11] While the topological organization of mechanisms is central to some mechanistic explanations (Levy and Bechtel 2013), this does not imply that all explanations that refer to topological properties are mechanistic (Brigandt, Green, and O'Malley 2018). I shall discuss the relationship between mechanistic and non-mechanistic explanation, including topological explanation, in Chapter 8.

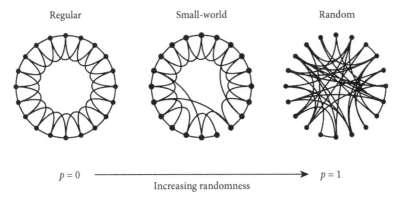

Figure 5.2 Watts and Strogatz (1998) on varieties of network organization.

including those in biological mechanisms (protein networks and neural networks), exhibit a small-world structure.   (Levy and Bechtel 2013, 249)

The explanatory relevance of this kind of network analysis depends upon the correctness of the claim that biological mechanisms are relevantly similar to these models in their network structure. But supposing that these similarity claims are warranted, this shows that mechanisms with diverse material entities and activities have common organizational features, which make them of a certain organizational kind (e.g., they are small-world networks). This in turn allows one to make general claims about these kinds (e.g., about the rate of propagation of signals through networks with small-world structures or the robustness of networks to damage to nodes or links) (Brigandt, Green, and O'Malley 2018).

Basic topological structure simply characterizes which nodes of a network are connected to which, but the organization of network structure may be more richly described by characterizing the functional relationships between nodes. For instance, are the networks Boolean networks in which nodes can be either on or off, are they connectionist networks with excitatory and inhibitory relations, or are they stochastic networks? What are the mathematical properties of relations between variables in the network? Are they, for instance, linear or non-linear? Also, how do the causal dependencies play out through time? Are there continuous interactive relations of the kind that can be represented by differential equations, or does propagation of influence through a network proceed in discrete stages? The range of possible functional relationships between parts of a mechanism is analogous to the range of functional relations among inputs and outputs of mechanisms that were discussed in Section 5.1.4. Indeed, in virtue of the mechanism dependence of activities and interactions, interactions between parts of a higher-level mechanism typically will just be activities of lower-level mechanisms.

5.2.5 Modularity

One last characteristic of the structure of mechanisms that should be mentioned is their *degree of modularity*. Intuitively, modular systems are those that can be broken down into subsystems or modules that play identifiable roles in the production of the phenomena, where the modules have reasonably isolable functional characteristics and have clear and limited connections to other modules. Modularity, as I am understanding it, is closely related to Simon's notion of near decomposability (cf. Wimsatt 1972) and thought to be an important characteristic of many biological systems (Wagner, Pavlicev, and Cheverud 2007). Simon says of systems that they are nearly decomposable when the strengths of interactions within subsystems is substantially greater than interactions between subsystems. According to Simon, such systems will tend to have the following properties:

(1) In a nearly decomposable system the short-run behavior of each of the component subsystems is approximately independent of the short-run behavior of the other components; (2) in the long run the behavior of any one of the components depends in only an aggregate way on the behavior of the other components.   (Simon 1996, 198)

Evidence of near-decomposability may be found in the topology of certain networks, with network neighborhoods corresponding to nearly decomposable modules, though it should be emphasized that whether a system is nearly decomposable depends not so much upon the existence of relationships to things outside a module, but the strength of those relationships.

The notion of modularity to which I am appealing is somewhat different from that found in the account of modularity that has received the most treatment in the philosophy literature over the last decade, originating with the work of Hausman and Woodward.[12] For Hausman and Woodward, modular systems are systems where it is possible to intervene in one part of the system without upsetting the causal relationships of that part to other parts downstream. As such, their notion of modularity is closely tied to the concept of an ideal intervention. As I understand it, modularity in the sense of near-decomposability provides an explanation of why it will be easier to intervene in such a way in a system. Modularity in this sense comes in degrees.

5.3 Types of Etiology

All mechanistic systems have histories. If a mechanistic system exists, it came to exist and have the capacities it has through the work of some causal process. This causal process is itself a mechanism—an etiological mechanism. Given that all mechanistic systems came from somewhere, another way of classifying mechanisms is by the kind of etiological mechanism that is responsible for the existence and capacities of the system.

---

[12] For discussion and criticism see Hausman and Woodward (1999; 2004) and Steel (2006).

One reason that it is important to distinguish the historical origin of a mechanism from its present constitution, organization, and capacities, is that it is possible to treat these aspects of a mechanism independently. As a case in point, consider the vexing issue of functions in biological systems. There are, broadly speaking, two ways to think about biological function—one concerns regular operation and causal roles within systems and the other concerns their causal origin. It is possible to characterize the function of a mechanism in the first sense, even if that function is not a "proper function" in the etiological sense. To ask questions about whether a capacity of a system has a proper function is to ask something about its etiology, rather than its current causal role. While we can distinguish between how the mechanism works now from how it came to be, traces of the historical origins of a mechanism are typically discernable in its present operation.

Certain terminological complexities arise from the fact that the processes that give rise to mechanical systems (the mechanism's etiology) are themselves mechanistic processes (the etiological mechanism). In what follows, I will use past participles to attribute to a mechanism a certain etiology (e.g., "evolved mechanism" or "selected mechanism") and will use adjectives or adjectival noun phrases (e.g., "evolutionary mechanisms" or "selection mechanisms") to refer to the etiological mechanisms themselves.

5.3.1 Designed and Built Mechanisms

One way that a mechanism can come to be is that some agent or agents design it and then build it. This is how machines like cotton gins and steam engines come to be, and, more generally, this is the etiology of most products of human engineering. An important feature of this sort of etiology is a clear separation between the designing and building of the mechanism and the mechanism's operation. A car is first designed by engineers and built by factory workers, and only then operated by drivers. Because of this, in the case of designed and built mechanisms, we can easily distinguish the mechanism for construction of the system from the mechanisms involved in the operation of the system.[13]

While our modern conception of mechanism has its origin in reflection on machines with design-and-build etiologies, mechanisms with this sort of etiology are a rarity in nature. Indeed, one might say that to call a mechanism a natural one is precisely to deny that it has a design-and-build origin. Design-and-build mechanisms ultimately require agents with cognitive capacities, and mechanisms of any complexity will require culture and social organization.[14]

---

[13] I am lumping together designing and building into a single category of etiology, though we could split them apart. There are some mechanisms that are built by other mechanisms, but where the mechanism doing the building was not designed but evolved. Proteins (which we wouldn't call mechanisms, but which are structured entities with mechanism-dependent capacities that serve as parts of mechanisms) are constructed by cellular "factories" (the mechanisms of protein synthesis), though these factories are the product of evolution rather than deliberate design.

[14] Designed-and-built mechanisms are exemplars of what Varela and Maturana (1980) call allopoietic systems. They contrast these with autopoietic systems, which they take to be autonomous, self-creating and

5.3.2 Evolved Mechanisms

Among mechanisms in nature, design-and-build is the exception, not the rule. The more common way for mechanical systems to come to be is via some sort of evolutionary process. In calling a mechanical system the product of an evolutionary process in this most general sense, I am not restricting myself to evolution by natural selection or even to evolution in biotic systems. Many natural systems, from landscapes to stars, can be said to evolve. The key distinctions I want to draw between evolved mechanisms and designed-and-built mechanisms are:

- Evolved mechanisms do not require designers.
- Evolved mechanisms acquire their capacities gradually, so that there is no sharp boundary between the time in which the mechanism is constructed and the time in which it operates.

As an example of an evolved mechanism, consider ecosystems. The ecosystem, and the many populations and abiotic environmental factors of which it is composed, will have many mechanism-dependent capacities. For instance, an ecosystem will produce biomass at certain rates; nutrients will be cycled through organisms in the system; it will absorb and transform energy. A currently functioning ecosystem is not a product of design, and it certainly is not constructed before it starts to operate. Rather, ecosystems evolve, and they acquire and transform these various capacities over time.

On classical accounts of ecological succession, an environment without a functioning ecosystem (say, a region where previous populations were destroyed by forest fires or volcanic eruptions) will first be colonized by pioneer species. The conditions created by these pioneer species permit further colonization by succession species; changes in the kind and distribution of species over time will alter the dynamics of various ecosystem processes. While ecological succession is a process occurring on longer timescales than day-to-day ecosystem functioning, ecosystem function is ongoing throughout the succession.

Because evolved systems do not have designers, evolved systems will tend to exhibit some kind of self-organization and self-maintenance (see Section 5.3.4).

While I have highlighted the difference between evolved mechanisms and designed-and-built mechanisms, it bears noting that most design-and-build processes in engineering are in fact to some degree evolutionary. Thus, the fact that a system has a designer, and that there is a distinction between the build phase and the operating phase does not preclude the idea that the mechanistic system has in some sense an evolutionary history. In engineering, as in biological evolution, constructing systems

---

self-maintaining systems. In their view, autopoiesis is the distinctive mark of living systems, and it is in the view of some (Dupré and O'Malley 2007; Nicholson 2012; Dupré 2013) incommensurable with mechanistic ontology and explanatory strategies. Nothing, however, in the account of minimal mechanism excludes autopoietic systems from being mechanistic; they rather constitute a distinctive kind of mechanism. Such mechanisms fit within the broad category of self-organizing mechanisms discussed below.

is often a matter of repurposing existing components or adding new components to existing systems. And as in other evolved systems, products of evolutionary design will tend to show their histories in their construction, in large part through the process that Wimsatt (2007) calls generative entrenchment. Additionally, some kinds of engineering applications increasingly depend on broadly evolutionary processes. For example, it has become increasingly clear that successes in artificial intelligence and robotics will depend centrally upon building systems that can learn. The most successful chess-playing programs, for instance, are designed and built to learn from the games they watch or play, so that they gradually acquire greater ability to win chess matches. Systems that learn are evolutionary systems in the sense that their capacities are not designed and built by some designer, but are instead acquired gradually as the system operates.

5.3.3 Selected Mechanisms

In his classic formulation of "Darwin's principles," Richard Lewontin (1970) suggests that natural selection will occur whenever a population has three properties: variation in traits (phenotypic variation), differential rates of survival and reproduction depending on those traits (fitness differences), and resemblance between the traits of parents and offspring (heritability). Natural selection is a process of change in the distribution of traits in a population over time that occurs whenever that population exhibits heritable variation of fitness.

Selection is one kind of etiological mechanism—that is, it is one kind of mechanism by which (current) mechanistic systems come to have the capacities they do. Selected mechanisms are a subclass of evolved mechanisms in which the evolution occurs in a population that exhibits heritable variation of fitness. Note here the terminology: a selected mechanism is a mechanistic system that has evolved through an etiological mechanism of natural selection.[15]

There are two main reasons why not all evolved mechanisms are selected mechanisms. The first is that evolution in general does not require populations. We can often say of a single system, or a single trait of a system, that it has evolved. We may say of the US Supreme Court or Butler University or the Windows operating system that it has evolved, but evolution in this sense simply means there has been a gradual change in the capacities (and the mechanisms underlying them). The second is that populations may change in ways that are not caused by natural selection. Textbook accounts of evolutionary biology emphasize that natural selection, while perhaps the main cause of evolution in biological populations, is not the only one. Others standardly

---

[15] Since Skipper and Millstein (2005), there has been considerable debate about whether natural selection is a mechanism in the sense of the New Mechanists. While I would agree with Skipper and Millstein that some of the features of early characterizations of mechanisms seem to be inconsistent with what is going on in natural selection, I believe natural selection falls within the scope of minimal mechanism. For a recent survey of this debate, as well as an argument that natural selection should be seen as a (minimal) mechanism, and a discussion of whether other causes of biological evolution can be considered to be mechanisms, see DesAutels (2016; 2018).

mentioned are mutation, migration, and drift. Beyond these orthodox explanations, some have argued (e.g., Kauffman 1993; 1995) that biological evolution (and certainly pre-biotic evolution) depends crucially upon the phenomenon of self-organization.

At the same time, it should be emphasized that selection mechanisms are a kind of etiological mechanism that can be found in many places both inside and outside biology. Many sorts of populations can exhibit heritable variation of fitness; these may be populations of organisms, populations of populations (metapopulations), populations of molecules, and so on. Darwinian selection models or "selection type theories" have been applied to a great variety of things, from a variety of sorts of cultural evolution to evolution of brain function and selection of clonal antibodies in the immune system (Darden and Cain 1989; Hull, Langman, and Glenn 2001; Mesoudi, Whiten, and Laland 2006; Godfrey-Smith 2007). Neither do Darwin's principles (at least in Lewontin's formulation) restrict selection to "natural" selection. Populations of domesticated organisms with traits that are the result of selective breeding (i.e., artificial selection) will be the production of selection mechanisms in this sense.

As with design, there are many kinds of selection, and there are many fruitful ways one might further classify selection mechanisms. These classifications might distinguish different sorts of selection on the basis of the kinds of entities, activities, and organization in the selective process: What is being selected? What are the mechanisms of inheritance? What is the source of variation? What are the sorts of concrete interactions that give rise to fitness differences?

5.3.4 Self-Organizing Mechanisms

Many mechanistic systems come to be through a process of self-organization. Self-organization in its most general sense occurs when a system organizes spontaneously because of some dynamics that falls out of the local interactions the parts of the system have with each other. A classic example in cellular biology is the formation of lipid bilayers, which form membranes around cells and sub-cellular structures. Phospholipid molecules in an aqueous environment will spontaneously organize themselves into a two-layered sheet. The reason for this is that the molecules are polar, having at one end a hydrophilic head and at the other end two hydrophobic tails. Contrast this with the wall that a builder may construct when building a house. Here the bricks do not align themselves, but require the work of an external agent.

Self-organization, while central to the understanding of the development of living systems, is by no means confined to it (Keller 2008; 2009). Crystalline structures (e.g., minerals, ice, snowflakes) arise through self-organizing processes in physical systems, as do atmospheres and weather systems. Self-organization is also a key feature of various chemical systems, including the sorts of pre-biotic chemical systems that could have given rise to life (Moreno and Ruiz-Mirazo 2009). Self-organization also occurs in many social systems. For instance, economic markets appear to be largely self-organized systems. Mechanisms with etiologies involving self-organization (i.e., self-organized mechanisms) represent a subspecies of evolved mechanisms. Self-organization and natural selection are in some sense alternate accounts of how a certain mechanism might have evolved. Many evolved systems will, however, arise through an etiological mechanism involving both selection and self-organization.[16]

One particular class of self-organizing systems is biological systems that result from processes of development. Biological organisms in all of their variety arise and acquire their capacities through a process of development and growth. Organisms reproduce, creating small and undifferentiated things that grow into larger and more differentiated things with various mechanism-dependent capacities. The etiology of biological capacities depends upon mechanisms operating at two scales—proximal developmental mechanisms operating on the scale of the individual organism and more remote evolutionary mechanisms operating at the level of populations, species, and higher taxa.[17] Developmental processes are self-organized at least in the sense that the organism is not built by an outside agent, though clearly these processes draw on an ensemble of developmental resources, many of which are in the organism's environment.

A great deal of recent work in theoretical biology and philosophy of biology has been devoted to understanding the nature of developmental processes and their relation to evolutionary processes, especially under the rubrics of evolutionary-developmental biology and developmental systems theory (Oyama 1985; Griffiths and Gray 1994; Müller and Newman 2005). I cannot explore or take sides on these matters here. It is enough to observe that the way that many biological systems come into being and acquire the mechanism-dependent capacities they do is through a process of growth.

5.3.5 Ephemeral Mechanisms and Entrenchment

Our discussion of the etiology of mechanistic systems has focused on the variety of ways in which stable systems can come to be and to acquire their capacities. The etiological mechanisms that produce these systems, while sensitive to background conditions and subject to various kinds of breakdowns, will often reliably produce mechanistic systems of certain kinds. On some occasions, however, the etiology of mechanistic systems may arise from conditions that are highly contingent and context-dependent. To the extent that they do, these systems are produced by ephemeral mechanisms (Section 2.3).

Mechanisms are collections of entities whose activities and interactions, organized as they are, are responsible for some phenomenon. An ephemeral mechanism is one in

---

[16] Much of the discussion of self-organization and its role in evolution can be traced to Kauffman (1993). For critical discussions and arguments for complementary roles for these two kinds of processes, see Weber and Depew (1996); Richardson (2001); Wimsatt (2001).

[17] Mayr's (1961) distinction between ultimate and proximal causes tracks fairly closely these two different ways in which we may describe the origin of an organism's capacities.

which the various parts, and with them the resulting activities and interactions, are "thrown together." More specifically, in an ephemeral mechanism:

- The configuration of parts is the product of chance or exogenous factors.
- The configuration of parts is short-lived and non-stable, and is not an instance of a multiply-realized type.[18]

The phenomena produced by ephemeral mechanisms are by their nature one-off events. We have already seen an instance in the example of the Danube flood. The particular characteristics of the flood, its timing and its extent, are dependent upon a variety of highly contingent and non-repeatable factors. It is still a mechanism, however, since, given the way that the parts of the mechanism are thrown together, the arrangement and dispositions of those parts will robustly produce certain outcomes. So, for instance, it may be highly contingent that there is a certain pattern of rainstorms in the Tyrol, but it is not contingent (given the characteristics of the ground and the topography of the region) that these rainstorms will lead to water flows of certain volumes into the Danube, water flows that will in the particular case produce the particular characteristics of the one-off flood event.[19]

One should not assume that, because a thing's origin is ephemeral, its ongoing behavior will also be ephemeral. Many one-off events produce systems that are responsible for recurrent phenomena. As an example of this, consider Old Faithful, the famous geyser in Yellowstone National Park. Old Faithful epitomizes a mechanistic system that produces a regularly recurrent phenomenon—namely eruptions in which hot water and steam are expelled from the earth in geysers of an average of 40 meters for durations of 1.5 to 5 minutes. The mechanism responsible for this consists, roughly speaking, of a chamber where water collects, a water source entering into the chamber, a heat source that heats the chamber, and a constricted vent that will, when the water reaches a critical point, release steam and water into the air. There are entities, activities, organization, and they produce a phenomenon. It is an accident of geological history—an ephemeral mechanism—that threw together these parts, but once put together, they maintain a stable system that has over recent history been producing eruptions faithfully around seventeen times per day.

Ephemeral mechanisms can under suitable circumstances give rise not just to single mechanistic systems that exhibit recurrent behavior, but to large classes of mechanistic systems whose characteristics mark their contingent origins. Stephen Jay Gould (1990) has famously argued that the pruning of the tree of life after the time of the Cambrian explosion seems to have depended upon highly contingent facts of history. The

---

[18] These two conditions come from Glennan (2010b), along with a third, which indicates that interactions "can be characterized by direct, invariant, change-relating generalizations." I am here dropping this as a necessary condition, in line with modifications to my treatment of interactions (see Chapter 2).

[19] It bears repeating that whether a process is taken to be ephemeral is a consequence of explanatory grain: the exact flood is ephemeral, but the grosser pattern of annual flooding is not.

argument has been generalized by John Beatty (1995), with his thesis that essentially all the distinctive features of biological systems are consequences of contingent histories. We need not enter into the debate about the extent that the features of living systems are the result of contingent accidents of history versus the optimizing power of selection (Gould and Lewontin 1979). It is enough to observe that features of biological systems depend at least in part on ephemeral mechanisms.

One important aspect of the relationship between ephemeral etiological mechanisms and robust patterns of behavior in mechanistic systems can be explained by the phenomenon that Wimsatt has called generative entrenchment. Generative entrenchment is, for Wimsatt, the historical process by which the contingent becomes necessary. Wimsatt writes:

A deeply generatively entrenched feature of a structure is one that has many other things depending on it because it has played a role in generating them. It is an inevitable characteristic of evolved systems of all kinds—biological, cognitive, or cultural—that different elements of the system show differential entrenchment.   (2007, 133–4)

Wimsatt's idea has its origins in thinking about developmental processes in organisms. Organismic traits that are fixed early in development become entrenched, quite apart from their inherent adaptive value, because later (downstream) developmental features depend upon them. Accordingly, mutations that affect these entrenched traits are more likely to be deleterious, as they will impact these downstream traits.

Wimsatt argues that generative entrenchment is a kind of historical phenomenon that can be seen across a wide variety of evolved systems—including cognitive and cultural systems. As an example of this, consider the mechanisms that gave rise to the current two-party system in the United States, and the ways in which they have become entrenched. Political scientists have observed that two-party systems arise from certain kinds of election systems, specifically those in which seats are distributed into single-member districts (with no proportional representation) and by "first-past-the-post" elections where the winner is determined by a plurality in a single ballot. The fact that US electoral politics has this structure is in large part a consequence of rules laid down in the US Constitution, a document whose historical origins are certainly to some degree ephemeral.

But even though its origins are ephemeral, the US two-party system has become deeply entrenched. The last major party reorganization in US history occurred immediately prior to the American Civil War, with the rise of the Republican Party as an anti-slavery party—and efforts to create third parties or replace one or the other party have proved fruitless, despite widespread voter dissatisfaction. Statutes, rules of parliamentary procedure, and informal conventions all institutionalize the Republican/Democrat structure. Because these institutionally entrenched rules govern the operations of the very mechanisms that could produce change, the institutions become deeply resistant to change.

5.4 Stochastic Mechanisms

Much of our thinking about mechanisms comes from thinking about machines, and many machines behave with "always or for the most part" regularity. Their behavior can be characterized by universal generalizations like "whenever you turn the key, the engine starts." These generalizations are not truly exceptionless of course; they are true, ceteris paribus. But when they fail to hold we take the reason to be that the mechanism is somehow broken, or some background condition fails to hold. The car is out of gas, the battery is dead, etc. Analogous ceteris paribus universal generalizations describe the behavior of mechanisms in nature: whenever the membrane depolarizes the action potential is triggered, or whenever the light dims our pupils dilate. Mechanisms of this kind can be said to be deterministic.[20] Many mechanisms, however, behave in ways that are not aptly described by ceteris paribus universal generalizations. Such mechanisms may be better described by statistical generalizations—generalizations that suggest that a mechanism behaves in a certain way only with a certain probability. For instance, we know that sexual intercourse among humans does not always lead to pregnancy, and that when it does, the probability of having an offspring of one biological sex or another is about 50 %. Mechanisms of this kind can be said to be stochastic.

These examples of stochastic and deterministic mechanisms emphasize the stochastic or deterministic *behavior* of those mechanisms. That is to say, these are examples in which the mechanism's phenomenon is stochastic. But this is not the only way in which mechanisms can be stochastic. Probabilities can play a number of different roles in characterizing aspects of a mechanism—its entities, activities, organization, and behavior—and these roles will lead to a variety of senses in which a mechanism can be stochastic.

To get a clearer sense of what stochasticity amounts to, it will be helpful to apply the models-first strategy. Random variables, probability distributions, and the like are mathematical entities that can be used in the construction of models of mechanisms, so to classify a mechanism as stochastic is in the first instance to classify it by the kind of model used to represent it, and reflection on the different roles such entities play in these models will help clarify the differing ways in which mechanisms can be stochastic.

Probability is a mathematical theory that can be characterized axiomatically by, e.g., the Kolmogorov axioms. Mathematical expressions in probability are open to various interpretations, both subjective interpretations in which probabilities measure something about evidence or rational belief, and objective ones in which probabilities are taken to describe some properties of systems, processes, or events in the world. Given

---

[20] "Determinism" is a convenient term here, but my use of it should not be taken to imply anything about larger metaphysical questions about determinism. I am using the term "deterministic" to refer to phenomena aptly characterized by universal rather than statistical generalizations. But the aptness of the description should be understood to be consistent with the idealized and ceteris paribus character of such generalizations. So, for instance, a watch would count as a deterministic mechanism, even if it sometimes loses time, and in erratic ways, and even though it stops working when the battery runs out.

that we are classifying mechanisms, rather than our beliefs about mechanisms, the relevant probabilities for our taxonomy will be objective in one way or the other.

Objective probabilities are connected to outcomes of chance setups. In a chance setup, there is some kind of event or trial that leads to a range of different outcomes on different occasions; for instance, a die is cast and lands on one of its six faces, or a subject is given a drug and either does or does not experience relief from some symptom. Repeated trials using the chance setup will lead to a distribution of outcomes. Chance setups like these have a natural mechanistic interpretation. The chance setup is a mechanistic system, and trials using the chance setup are mechanistic processes. Processes set up in this way are examples of input-output mechanisms (Section 5.1.3). The input is whatever starts the trial (the tossing of the die or the taking of the drug) and the output is the trial's result. Because the trial does not turn out the same way on all occasions, the input-output behavior is described not by a universal generalization, but rather by a probability distribution.

While probabilities may be useful in describing some one-off mechanisms, our primary use of stochastic characterizations of mechanisms is for recurrent mechanisms. In such mechanisms, the chance setup occurs repeatedly, leading to different outcomes on different occasions. While there is perhaps one sense of the term "regular" where for a mechanism to be regular it must behave always or for the most part the same way on every occasion, stochastic mechanisms that do not have this property are still regular in important ways. Consider a mechanism that rolls a single six-sided die. Assuming the mechanism is fair, the die will have a probability of 1/6 of landing on each face, and over repeated trials the relative frequency of each of the outcomes will under most conditions approach 1/6. While the outcome of a trial is not the same on most or every occasion, the *distribution* of outcomes is highly regular.

The basic distinction between deterministic and stochastic types of mechanisms is between mechanisms with models that characterize their behavior deterministically (via universal generalizations) and mechanisms with models that characterize their behavior probabilistically (via statistical generalizations). But probabilities can show up in different places in the representation of a mechanism, suggesting different senses in which a mechanism can be stochastic. There are three main places where probabilities can be used to describe a mechanism:

- The mechanism's phenomenon can be characterized by probabilities.
- The likelihood that entities act or interact can be characterized by probabilities.
- The outcomes of activities and interactions can be characterized by probabilities.

The first kind of probability characterizes the mechanism as a whole, while the second and third characterize the activities of the parts. I distinguish between the second and third cases because there are both probabilities that activities/interactions occur and probabilities that characterize how these activities or interactions turn out. To see the difference, consider again the wolf-sheep model discussed in Chapter 4. Two of the activities the wolves engage in are moving about the landscape and eating sheep. Both of these activities could be described probabilistically. A given wolf might move by chance in one direction or another, and a given wolf who meets a sheep will have a certain chance of managing to eat it. The chance character of the paths of the wolves (and the sheep) means that there is only a certain probability that a wolf will meet a sheep at any given time. This is the probability that the wolf and sheep interact. And given that the wolf and sheep interact, there is another probability that characterizes the outcome of the interaction—does the wolf eat and the sheep die, or does the sheep get away?

What is striking about stochasticity in mechanisms is that it can emerge, disappear, and re-emerge at different levels of mechanistic organization. Collections of entities whose activities and interactions are stochastic may be responsible for phenomena that are approximately deterministic. For example, consider the phenomenon of chemotaxis—the process by which single-cellular organisms or cells move along chemical gradients. Describing the chemotaxis mechanism in *E. coli*, Abrams (2018) shows how repeated stochastic interactions of parts of the bacterium with molecules in its environment change the behavior of its flagella in ways that will move the bacterium toward greater concentrations of a desirable substance. While at high spatial and temporal resolutions the bacterium's path is random and not unidirectional, the path is biased so that the overall direction of the movement is deterministically along the chemical gradient.

Conversely it is possible for systems whose parts behave in deterministic ways to generate stochastic behavior. Paradigms of such systems are those used in games of chance. Abrams writes:

There are some devices . . . whose causal structure is such that it typically matters very little what pattern of inputs the device is given in repeated trials; the pattern of outputs is generally about the same. For example, a wheel of fortune with red and black wedges—a simplified roulette wheel—is a deterministic device: The angular velocity with which it is spun completely determines whether a red or black outcome will occur. Nevertheless, if the ratio of the size of each red wedge to that of its neighboring black wedge is the same all around the wheel, then over time such a device will generally produce about the same frequencies of red and black outcomes, no matter whether a croupier tends to give faster or slower spins of the wheel.

(Abrams 2012, 349)

The roulette wheel produces statistical patterns of recurrence from the repeated operation of a single mechanism, and for this reason, characterizing those patterns of inputs and outputs in terms of probabilities does not reflect ignorance. The roulette wheel thus illustrates the possibility that deterministic mechanisms may behave in objectively chancy ways. While the systems used in games of chance are obviously engineered to produce this kind of behavior, it seems likely that a variety of complex systems found in the physical, life, and social sciences have a similar character.[21]

---

[21] Just how to formulate a theory of these deterministic but objective probabilities is a matter of some dispute, but many authors argue that there are such probabilities and that they can be found in many scientific domains (Strevens 2005; 2011; Hoefer 2007; Sober 2010; Abrams 2012).

We should close our discussion of stochastic mechanisms with a note of caution. One concern about using a stochastic representation of the behavior of a class of mechanisms is that sometimes the variation in outcomes across cases may not be characterized by a similar stochastic mechanism, but may instead merely hide our ignorance of differences between mechanisms operating in particular cases (Glennan 1997a). The following example will illustrate the problem. The drug thalidomide was used in the late 1950s and early 1960s to treat nausea in pregnant mothers, before it was discovered that its use greatly increased the risk of severe birth defects. We can represent the relationship between thalidomide use and birth defects as the result of a stochastic input-output type mechanism, with the input being thalidomide (or not) and the output being the baby (healthy or not). To make the example very simple, suppose the probability of birth defects given ingestion of thalidomide in the first trimester is 10%, otherwise 0%. This pattern of behavior among thalidomide mothers is compatible with the multiple interpretations: One possibility is that the effect of thalidomide is the same on each mother, in each case raising an objective chance of birth defects to 10%. A second possibility is that thalidomide deterministically produces birth defects in children of mothers that carry a particular undetected genetic characteristic, one that happens to be carried by 10% of the population. If the first possibility obtains, the behavior of this mechanism is genuinely stochastic; if the second possibility obtains, the pattern of variation is simply the result of a population-level average over two different types of mechanisms. If the second case obtained, our treating the thalidomide birth defect mechanism as a single kind of stochastically behaving mechanism would simply be a mistake.

While it is appropriate to look for the mechanism responsible for a statistical pattern, it does not follow from the fact that there is a statistical pattern in a set of trials that each trial's outcome is produced by a similar mechanism. It is, for instance, the case that there are strong statistical associations between the income levels of fathers and sons, but investigation of this pattern will inevitably reveal enormous variation between individuals (and between subpopulations, like those in different counties) in how income levels of fathers influence those of sons. The production of this pattern can only be explained as the combined production of a population of heterogeneous mechanisms.

5.5 Some Mechanism Kinds

The strategy of mechanism taxonomy developed here leads to a cross-cutting system of classification. A token mechanism will be an instance of many different types, and mechanisms that are heterogeneous with respect to some features will be clustered together with respect to others. An alternative approach to classification is to identify a mechanism kind by a cluster of features that are typical of members of that kind. Some examples of mechanism kinds that could be described in this way are market mechanisms, genetic mechanisms, and metabolic mechanisms. Mechanisms of these kinds may be characterized by their structural as well as material similarities, by their characteristic phenomena, organization, and etiology. Much of scientific inquiry is organized around developing accounts of the structure and behavior of these kinds of mechanisms.

In this section I will show how mechanism kinds of this sort can be described using the features I have identified in this and the previous chapter. I have chosen three examples—machines, computational mechanisms, and social mechanisms—which can help illuminate the variety there is to be found in mechanism kinds.

5.5.1 Machines

New Mechanists have rightfully advised against identifying mechanisms with machines, but on the account of minimal mechanisms, it is possible to characterize machines as a certain kind of mechanistic system. Like the term "mechanism," however, the term "machine" has a long history and no clear boundaries. This means that we cannot have a general model that clearly identifies the features of machines in general. What we can do, however, is to use the various taxonomic features that we have identified in this chapter to describe a cluster of features typically associated with machines, and thereby to explain how they are different from other kinds of mechanisms.

The etymology of the word "machine" tells us something immediately about one of its most essential features. It comes from a family of ancient Greek words meaning device or contrivance. In short, machines are artifacts, constructed so as to serve some purpose. In terms of our taxonomy, machines are typically examples of designed-and-built mechanisms.

While prehistoric tools also have designed-and-built etiology, we can think of the birth of machines as contemporaneous with the development of large organized societies in the historical period. Ancient civilizations, from China to Egypt and the ancient near east, constructed machines to help them build and maintain their material environment, as well as to engage in warfare—machines like cranes, carts, and catapults. The birth of machines in antiquity is really the birth of mechanical engineering, for it is engineers that design and build the machines.[22]

While machines can produce, underlie, or sustain a wide variety of phenomena, there are a number of features that are most often associated with them. First, the phenomena for which machines are responsible are typically recurrent. Machines are designed to be used more than once. A catapult, for instance, will repeatedly throw a stone approximately the same distance, given the same size and shape of stone. With recurrence also comes the idea that machines have (normatively) normal functions, and the capacity to malfunction.

Another way to identify machines as a kind of mechanism is by considering the kinds of entities that are machine parts, and by the kinds of activities and interactions in which they engage. Machines of the kind built by premodern engineers typically have so-called "simple machines" as basic parts. These simple machines, still discussed

---

[22] Bautista Paz et al. (2010) provides a helpful history of machines from the perspective of mechanical engineering.

in engineering textbooks, are parts or small collections of parts that translate or magnify force and power. Standard simple machines include the lever, the screw, the pulley, the wheel and axle, the wedge, the inclined plane, and gears. These kinds of parts can be materially different, but they must have certain properties to function as parts of these kinds. Most importantly, they must have appropriate strength and rigidity. A lever, for instance, must not bend upon application of force. A classical complex machine is an organized collection of these simple machines. For instance, a water-powered mill will contain a series of parts that translate the lateral motion of water in a stream into the circular motion of the millstone. Advances in the physical sciences and technologies permit the construction of machines using different kinds of parts that interact with each other in different ways. In addition to the pushes and pulls of "strictly mechanical forces," machines can be built that use electromagnetic, chemical, and optical components and interactions, to name just a few possibilities.

But while many contemporary machines and devices are materially distinct from their historical predecessors, they share many structural and organizational features with these early machines. In terms of the organizational features identified in Section 5.2, we can say that machines tend to be few-parted, that their parts tend to be functionally differentiated, that their organization tends to be induced rather than affinitive, and that their construction tends to be modular. All of these organizational features come in degrees, and certainly there are considerable variations in these features among different classes of machines. But even so, these are common tendencies, and the reasons are not difficult to see. These organizational principles tend to be practically good engineering principles. A modular system, for instance, is easier to build, maintain, and repair.

Machines are probably too amorphous a class to count as a genuine kind. Certainly, this is the case if we stick with the models-first approach to kinds, since there is no general model of machines as such. What we do have is lots of classes of machines that share a cluster of properties—a cluster that shows both that machines are mechanical systems and that there are many mechanical systems that are not machines.

5.5.2 Computers and Computing Mechanisms

In a series of papers (Piccinini 2007a; 2007b; 2008; 2010; Piccinini and Bahar 2013), Gualtiero Piccinini and his colleagues have developed a theory of computers and computation that treats computers and other computational systems as kinds of mechanisms. Using this account, we can show how computers and computing mechanisms fit within our broader taxonomy of mechanism types.

On Piccinini's account, computing mechanisms are mechanisms that perform computations, and computations are certain kinds of transformations of inputs to outputs. Piccinini and Bahar define a generic notion of computation as follows:

Computation in the generic sense is the processing of vehicles (defined as entities or variables that can change state) in accordance with rules that are sensitive to certain vehicle properties and, specifically, to differences between different portions (i.e., spatiotemporal parts) of the vehicles. A rule in the present sense is just a map from inputs to outputs; it need not be represented within the computing system. Processing is performed by a functionally organized mechanism, that is, a mechanism whose components are functionally organized to process their vehicles in accordance with the relevant rules. Thus, if the mechanism malfunctions, a miscomputation occurs.   (Piccinini and Bahar 2013, 458)

On Piccinini and Bahar's account, computing mechanisms are defined in the first instance by the kinds of phenomena they produce. Computing mechanisms are a species of input-output mechanisms that exhibit recurrent behavior. That recurrent behavior is represented in the rules by which the vehicles are processed. Computing mechanisms have normal behavior in both the statistical and normative sense—normative because miscomputations can occur.

Piccinini and Bahar offer a conception of computation that is generic, because they want to distinguish three subspecies of computation that can fall under it, namely digital, analog, and neural computations. I shall not discuss the analog and neural varieties of computation, and Piccinini and Bahar's contention that neural computation is a distinctive kind, but will confine myself to the digital variety. According to Piccinini and Bahar, digital computations are computations in which the vehicles are strings of digits, and in which the processing involves rule-based transformations of one string of digits into another. Digits are understood to be entities in or states of a system that form a finite alphabet, for instance, "T" and "F," "1" and "0," or the letters of the Roman alphabet.

A computational mechanism, as any mechanism, is a concrete thing located in space and time. In the case of digital computation, Piccinini and Bahar suggest that concrete digits have the following properties.

[A] digit is a macroscopic state (of a component of the system) whose type can be reliably and unambiguously distinguished by the system from other macroscopic types. To each (macroscopic) digit type, there correspond a large number of possible microscopic states. For instance, a huge number of distinct arrangements of electrons (microscopic states) correspond to the same charge stored in a capacitor (macroscopic state). Artificial digital systems are engineered so as to treat all those microscopic states in one way—the way that corresponds to their (macroscopic) digit type.   (Piccinini and Bahar 2013, 459)

A digital computational mechanism is a system whose parts are organized in such a way as to transform strings of concrete digits into other strings of digits in accordance with the abstract rules that describe the computation. Those entities will include bottom-out computational processors—logic gates—whose operations correspond to the transformation of input digits to output digits according to specific rules (e.g., conjunction, disjunction).

One of Piccinini's main goals in offering an account of computational mechanisms is to avoid what he calls "computational nihilism" (Piccinini 2007b). It is sometimes asserted (especially in discussions of computational theories of cognition) that computation is in the eye of the beholder, and more or less any systems (e.g., weather systems or stomachs) can be interpreted as performing computations. A common reason given for this view is that any physical system will undergo transitions in physical state, and it should be possible in principle to create a mapping between these state transitions and stages in an abstract mathematical computation, so according to that mapping, the system is seen as performing the computation.

It seems hard to deny that such mappings would be possible in principle, but they should not worry us. On such a weak and unconstrained notion of computational description, more or less anything could compute more or less any function. But our practice of forming mathematical descriptions of physical systems, both natural and engineered, suggests that there are serious constraints on what kinds of systems can be seen as performing digital computations. This starts with the notion of a digit itself. Physical systems that can perform digital computations must be able to store digits so that those digits persist, other components must be able to read those digits and change them, and so on. This requires that components representing digits must be able to switch between a finite number of discrete and stable states (two states—on and off—for a Boolean computer, but possibly more). Only some kinds of components can do this. To put the matter metaphorically, only some kinds of objects look like switches; others look like dials, and others cannot be set to a value at all.

A more plausible argument in support of a computational nihilist view points to the possibility of computational models for all sorts of physical systems. It is common practice to use computers to create computational models of physical systems like digesting stomachs or hurricanes. These models are clearly performing computations, so why then are the things they are modeling not performing computations as well?

Piccinini's response to this objection is to make a distinction between computational modeling and computational explanation (Piccinini 2007b). Even if all systems have computational models, systems will have computational explanations only if it is the function of those systems to perform computations. Function in this sense is a teleological notion, meaning that the system must execute computations either as a matter of design or selection. For a mechanism to be a computational one, it must have a certain kind of etiology. It is only because of this etiology that it makes sense to speak of computational malfunction. If we look at a system like a hurricane, which can be modeled computationally but which does not actually perform computation, it is not possible for it to miscompute.

It might seem that Piccinini's defense of the reality and objectivity of the distinction between computational mechanisms and other kinds of systems is incompatible with my weakly realist models-first approach to kinds, for it might seem like on the models-first approach, the existence of the computational model is enough to make something a computational kind. But this is not really the case. For something to be a computational mechanism it must have a certain kind of etiology; it must be designed-and-built to perform computations, or it must have evolved via selective processes to perform computations. And the sense of "to perform computations" implies something about how the computational mechanism is embedded within its larger context.

To the extent that there is a difference between Piccinini's view and the models-first approach, it is in the fact that the models-first approach suggests that there is no account of whether something is a mechanism of any kind in the absence of a model. But we can reconcile Piccinini's views with the models-first approach by observing that there are only a limited class of systems which can appropriately be modeled as having the kind of etiology and embedding required of computational mechanisms. In the style of Daniel Dennett's very modest realism, we make this point by saying that there are only a limited class of systems where it is fruitful to take a computational stance.

5.5.3 Social Mechanisms

Around the same time philosophers of science began to talk about mechanisms in the late 1980s and early 1990s, a parallel but largely independent discussion was occurring among some theorists in the social sciences. The motivations for both discussions were similar, including a dissatisfaction with logical empiricist accounts of explanation and with the explanatory poverty of correlational data, as well as a sense that genuine explanations of complex systems required understanding the interactions between and organization of their parts. In the case of the social sciences, the mechanistic approach is closely connected with the doctrine of methodological individualism, which is roughly the idea that any social phenomenon is ultimately caused by, and therefore must be explained in terms of, the beliefs, intentions, and actions of agents.[23]

While the term "mechanism" has wide use in the social sciences, there is no scholarly consensus on exactly how to characterize them. James Mahoney (2001) in a review of work on mechanisms in sociology found no less than twenty-four distinct definitions or explanatory glosses on the concept. Many of these variations, however, may be of little theoretical consequence, and these definitions clearly are motivated by observations and intuitions similar to those of New Mechanists. A review paper about social mechanisms (Hedström and Ylikoski 2010) identifies nine definitions from the recent literature—four from the New Mechanists and five from social science theorists. They argue that there is much overlap between these definitions, and I would suggest that many of the shared features are captured by minimal mechanism.

The definitions that Hedström and Ylikoski review characterize mechanisms in general, rather than social mechanisms in particular, but the question I want to consider here is what distinguishes social mechanisms as a kind from other sorts of mechanisms. We can use our catalog of mechanism features to help answer this question.

Social mechanisms are in the first place defined materially—by the kinds of entities and activities that make them up. The ultimate constituents of social mechanism will always include active agents. Typically the agents that social scientists are interested in are human beings, but social phenomena can be produced by mechanisms in which

---

[23] Jon Elster (1989; 2007) was a pioneer of this approach, as were Hedström and Swedberg (1996). Other useful sources include articles in Hedström and Swedberg (1998), as well as Bunge (1997); Little (2011); Demeulenaere (2011); Steel (2005); and Hedström and Ylikoski (2010).

the agents are non-human. What is required is agency as such. I will not try to say what is required for agency, and clearly agency admits of degrees, but at the least to be an agent in a social mechanism, something must have beliefs, desires, intentions, and the ability to causally interact in the world on the basis of these. Agency requires a certain amount of cognitive capacity, but clearly there exist social phenomena that are produced by the collective behavior of non-human animals—baboons, elephants, dolphins, dogs, and so on. If agents are the primary entities in social mechanisms, the primary activities and interactions are the ones in which these agents engage—forming friendships, antagonisms, romantic and family bonds; buying and selling, eating together, raising children, and so on. Social mechanisms will involve interactions not just among agents, but also between agents and elements of their natural and socially constructed environment. Economic mechanisms, for instance, are a species of social mechanism, but economic phenomena will involve the exchange of goods and services, and those goods and services will be or use material entities that are not agents.

The other prominent feature of social mechanisms is that they typically explain the relationship between these individual agents and larger collectives—families, communities, social institutions, nations, etc. Hedström and Ylikoski write:

For sociology, the most important explananda are social phenomena. Although sociology shares with psychology and social psychology the interest in explaining properties and behaviors of individuals, its key challenge is to account for collective phenomena that are not definable by reference to any single member of the collectivity. Among such properties are…

1. Typical actions, beliefs, or desires among the members of society or a collectivity.
2. Distributions and aggregate patterns such as spatial distributions and inequalities.
3. Topologies of networks that describe relationships between members of a collectivity.
4. Informal rules or social norms that constrain the actions of the members of a collectivity.

(Hedström and Ylikoski 2010, 58-9)

While kinds of phenomena listed here may be of special interest to sociologists, the interdependence between parts and wholes is characteristic of phenomena that are studied by social scientists in a variety of disciplines, including anthropology, economics, and political science. The kinds of phenomena exhibited by social mechanisms are typically constitutive; certainly this is the case with phenomena of types two and three. Social phenomena are also typically represented stochastically, in terms of probabilistic associations between independent and dependent variables.

Hedström and Swedberg (1996; 1998) conceptualize the causal processes involved in changes at collective (or macro) levels as involving three types of mechanisms, which they call situational mechanisms, individual action mechanisms, and transformational mechanisms. They represent the relationship between these three sorts of mechanisms using a diagram (Figure 5.3), originally from Coleman (1986), often called a Coleman boat, with the three sides of the boat representing the three kinds of mechanisms.

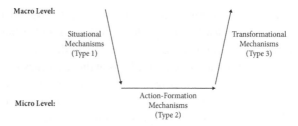

Figure 5.3 A typology of social mechanisms. Redrawn from Hedström and Swedberg (1998).

In situational mechanisms, agents are influenced by collectives and their properties; in action-formation mechanisms, individuals act and interact with other individuals; and in transformational mechanisms, the aggregated behavior of individuals brings about changes in the collective. As an example of this sort of pattern (again following Coleman), Hedström and Swedberg (1996, 298) cite Weber's account of the role of the Protestant ethic, in which Protestant religious doctrines influence individual beliefs, which affect individual values, which in turn generate a set of economic behaviors, which collectively lead to the growth of capitalism.

While Hedström and Swedberg's three kinds of social mechanisms (represented by the arrows) each count as mechanisms on the minimal mechanisms account, so too does the diagram in its entirety. Mechanisms are individuated by their phenomena, and there are multiple phenomena described here. There are first of all the phenomena marked by the individual arrows, for instance the mechanism by which religious doctrines influence individual actors' beliefs. But the macro-level association is also a phenomenon, in this example the association between the development of Protestant theological doctrines and the growth of capitalism. Indeed, looking for the mechanisms responsible for these macro-level associations is arguably one of the major goals of sociology and the social sciences more generally.

Social phenomena depend upon a combination of causal and constitutive relationships, so care needs to be taken in interpreting the arrows within the diagram. At first sight, the arrows look like causal arrows, with the first representing "top-down" causation and the third "bottom-up" causation. We should be careful, however, not to say that these individual activities produce collective phenomena. Rather, it is the case that the collective changes to individual actors constitute (or "just are") the change in the collective. It is an example of what Craver and Bechtel (2007) call a "mechanistically mediated effect" (see Section 7.9). Similarly, it is not the case that a property of the collective (like commonly held religious doctrines) *produces* changes in the individual. Protestant theological doctrine is neither an entity nor a property of an entity that can act upon an individual agent. An individual person's values are ultimately produced by that person's particular interactions with her environment. It is not the Protestant ethic as such, but the particular sermon, the repeated prayers at the dinner table, or the particular things that Mom, Dad, and others say that shape the particular individual's dispositions. Still, there is an important sense in which properties of collectives make a difference to the behavior of individuals. How to understand causation in such systems will be addressed in the next two chapters.

5.6 Conclusion: Types of Mechanisms and the Unity of Science

The search for types of mechanisms is at root an expression of the scientist's craving for generality. For even though mechanisms are particulars, it is only by achieving general knowledge about kinds of mechanisms that scientists can describe, explain, predict, and control things in the world. The analysis I have offered in the last three chapters suggests how that generality can be found. It does not lie in an unrepresented world, but it can be constructed and discovered through the work of science—constructed because modeling and representation is a constructive activity, but discovered because those models, if they are to serve our purposes, must represent features of the world. The features scientists choose to attend to or ignore reflect their varying goals and epistemic resources, and different abstractions and idealizations lead to different ways of classifying things in the world.

The picture I have given of mechanism kinds gives us some understanding about the nature of disciplinarity and of the extent and limits of the unity of science. Scientific fields are largely defined by what I have called material similarities—similarities in what (material) kinds of phenomena they seek to explain, as well as the set of entities, activities, and interactions that they take to be responsible for these phenomena. Molecular genetics, for instance, is defined principally by certain phenomena it seeks to understand (e.g., how, when, and in what quantities are proteins synthesized) and by the entities and activities that are involved in the mechanisms that are responsible for these phenomena (e.g., DNA, RNA, proteins, transcription, translation, and regulation). Disciplines grow around the material and theoretical resources, technologies, and experimental techniques used to explore these phenomena and the mechanisms responsible for them.

But scientific fields are not islands. For one thing, they are integrated by what Darden and Maull (1977) once called interfield theories. This integration does not come via a grand theoretical reduction, but rather by exploring localized relations of mechanism dependence, where entities or activities assumed in one field are located, filled in, and explained in other fields (Bechtel 2009; Darden and Craver 2002). For another, scientists operating in seemingly unrelated fields can often benefit from tools that exploit structural features of mechanisms and phenomena. Techniques like dynamical modeling, agent-based modeling, and network analysis can help scientists from across the physical, life, and social sciences to understand phenomena that are produced by very different kinds of entities and activities, so long as those systems and processes have structural features that make such analyses possible.

The diversity of mechanisms explains why simple one-size-fits-all models of scientific method do not work. But the similarities that one finds, especially in the organization of mechanisms, account for interdisciplinary connections across the sciences. One is left with a picture of science that is pluralistic in its methods but in which different sciences are bound together in a single fabric, both by discipline-crossing methodologies and by the patterns of mechanism-dependence that connect the entities and activities studied by those disciplines.

6

Mechanisms and Causation

What is the relationship between mechanisms and causation? Put briefly, it is just that causes and effects must be connected by mechanisms. If one event causes a second, there exists a mechanism by which the first event contributes to the production of the second. If some kind of event causes another kind of event, it is because there are one or more kinds of mechanisms by which events of one kind cause events of the second. While this is the core of the account, much must be done to flesh it out.

While it is generally granted that mechanisms are central to at least some kinds of causal explanations, there has been considerable skepticism about whether it is possible to offer an informative mechanistic account of the nature of causality. As Max Kistler puts it, "on closer inspection, it appears that the concept of mechanism presupposes that of causation, far from being reducible to it" (Kistler 2009, 599). Similarly, Franklin-Hall argues, New Mechanists either must adopt an independent account of causation like Woodward's (2003) manipulability approach or appeal to a concept of activity that is, "from a philosophical perspective, brute" (Franklin-Hall 2016).[1] Even some prominent defenders of the New Mechanism as an approach to understanding explanation and discovery have expressed skepticism about a mechanistic account of causation.[2]

I will argue that this skepticism is not warranted. Part of the work has already begun with the discussion of activities and interactions, where I have shown that we can in fact say something informative about activities in general. I will continue this work in this chapter and the next by showing explicitly how one can use mechanisms to provide an explicit, albeit non-reductive, account of causation.

Much of the work of this chapter will be to situate a mechanistic approach within the landscape of the recent philosophical literature on causation. In particular, I will show how a mechanistic account of causation fits within the more general category of singularist approaches to causation, and I will discuss how it contrasts with regularity and difference-making approaches, especially Woodward's influential manipulability account. The bottom line of this discussion will be that these other accounts offer important insights into how causes are found and used, but that, ontologically, causal relations depend upon the existence of actual mechanisms. The point of the comparative

---

[1] For other criticisms see Williamson (2011); Psillos (2004); Casini (2016).

[2] In his 2007 book, Craver explicitly disavowed a mechanical approach to causation: "I do not think that causation can be explicated in terms of mechanisms, as [Glennan's] mechanical account claims, but I do believe that explanations often describe mechanisms" (Craver 2007, 86).

and polemical work of this chapter is to make intuitively plausible the viability of a mechanistic theory of causation. The account will be filled out in the following chapter, where I will offer a mechanistic analysis of the concepts of causal production and causal relevance and defend the theory against a number of plausible objections.

6.1 Methodological Preliminaries

The topic of causation can be approached from a number of directions: among them semantic, epistemic, and ontological (or metaphysical). A semantic approach is concerned with the meaning of causal concepts and claims; an epistemological approach starts from questions about how we come to know and understand causes; an ontological approach focuses on the features of the world that make our causal claims and explanations true. The mechanistic theory I will offer is meant to be an ontological account, but it cannot avoid touching on semantic and epistemic issues.

A number of philosophers have sought to explore the metaphysics of causation via Canberra planning (Menzies 1996; Lewis 2004). Canberra planning is a two-step process that begins with semantic analysis—offering an account of the meaning of causal claims by identifying platitudes about causation. The initial step identifies the role of causal concepts within our broader commonsense and scientific conceptual economy and sets the stage for a second step, whereby philosophers (informed by science) identify what features of the world occupy this role.

As was noted in Chapter 1, a major worry about the Canberra plan is that it may simply not be the case that the platitudes associated with a concept are sufficiently clear and consistent to pick out some single feature of or relationship in the world. This is certainly a concern for causation. It has been widely observed that rules for assigning causes of events appear to be heavily dependent upon pragmatic considerations. For instance, if someone dies of radiation poisoning, one typically counts the person's exposure to radiation as a cause of their death, but one does not typically count their birth as a cause of their death. This difference in usage appears not to rest on some ontological difference, but upon some set of human interests, explanatory, legal, or otherwise. More importantly, as will become clear through this chapter, there appears to be conflicting platitudes about causation, which lead to distinct and non-equivalent concepts of cause.

There is also a special problem with applying the Canberra plan to causation.[3] The general approach involves using conceptual analysis to understand the functional role of something, and then using empirical analysis to find out what fulfills this role. But what exactly is a functional role? In general it is a *causal role*! So, for instance, in Lewis's original deployment of the strategy (Lewis 1972), a mental state like pain is whatever it

---

[3] For another discussion of the values and hazards of Canberra planning with causation see Illari and Russo (2013, ch. 19). For a more general discussion about the pitfalls in separating conceptual and empirical analyses of causality, see Bontly (2006).

is that is caused by thus and such, and causes thus and such. How can we apply such an approach in a non-circular way? Perhaps there is a way to specify the folk conception of causation in a way that is independent of the "causal" role that causes play. For instance, the platitudes that causes are distinct from their effects or precede their effects are non-causal conditions on what it is to be a cause. But, even so, there is much evidence that our folk conception of causation is closely tied to what we can *do* with causes. This is the central intuition behind Jim Woodward's manipulability approach to causation (see also Menzies and Price 1993). In the very first two sentences of his influential book, Woodward writes:

This book defends what I call a manipulationist or interventionist account of explanation and causation. According to this account, causal and explanatory relationships are relationships that are potentially exploitable for purposes of manipulation and control.   (2003, v).

The idea that causal relationships are exploitable for manipulation and control is an important example of a platitude about causation, the sort of thing that is clearly part of what we mean by "cause." But it is evident that this "functional role" of causation is defined in causal terms, since manipulation and control are themselves causal concepts.

Whether this is a problem or not depends upon what you want, or at least what you think you can get, out of a theory of causation. Most ambitiously, you might be after a theory of causation that reduces causation to something else. This was Hume's aim, when he argued that causation is reducible to constant conjunction: On his account, what makes it true that A caused B is that all A-like things are followed by B-like things. The reduction was both semantic and ontological. Hume argued both that all we can mean by cause is constant conjunction, and that the features of the world upon which the truth of causal claims depend are patterns of constant conjunctions.

But analysis need not be reductive to be informative. A non-reductive analysis like Woodward's does not seek to define causes in terms of something else, but instead seeks to show the interrelationship between various kinds of causal concepts. For Woodward, this is a matter of showing the relationship between causation in general and a certain specific sort of causal relation—manipulation. The mechanistic account of causation that I will be offering will, like Woodward's, be conceptually non-reductive. It cannot conceptually eliminate causes, since the concept of mechanism to which the mechanistic account will appeal is itself a causal concept. In contrast to Woodward, however, my focus will be ontological rather than conceptual and epistemic. In the Canberra style, I will use the conceptual analysis as a stepping-off point, but my ultimate concern will be with causation as a relation in the world.[4]

The most familiar approach to ontological questions about causality is to offer an account of "the nature of the causal relation." Such an account will characterize

---

[4] See Woodward (2003, sec. 1.7) on the virtues of non-reductive analysis. See Steel (2008, 47) for a defense of including manipulability requirements in Canberra-style analyses of causation.

both the relata (what kinds of things can be causes or effects) and the relation (what conditions are required for cause and effect to be related). In my view, the problem with this approach is that it may lead to an overly simplified view of the ontological grounding of causal claims. It assumes that there is one thing that is *the* causal relation and that this relation ultimately holds between relata of some particular ontological category (facts, events, etc.).

The New Mechanist account of activities and interactions suggests a different tack. According to this account, activity (and with it cause) is an abstract concept that is used to characterize the actual concrete activities and interactions that are productive of changes in the world. There is on this view no one thing which is interacting or causing, and when we characterize something as a cause, we are not attributing to it a particular role in a particular relation, but only saying that there is some productive mechanism, consisting of a variety of concrete activities and interactions among entities. The totality of mechanisms—including their (generally mechanism-dependent) parts, activities, and interactions—constitutes "the causal structure of the world."

If this approach is correct, then much of the ontological account of "the nature of causation" has already been given implicitly in our account of activities, interactions, and mechanism-dependence. But work still remains. To start, we must explore the relationship between causal claims and the mechanisms upon which the truth of those causal claims depend. There are, as we shall see, a variety of different claims, and these different kinds of claims will depend in different ways upon mechanisms.

6.2 The Variety of Causal Claims

Causal claims come in many forms; they make different kinds of assertions about the causal-mechanical structure of the world, and hence the truth-makers for these different claims will be of different kinds. While it is not possible to offer an exhaustive account of the varieties of causal claims, we can identify some major forms.

The obvious place to start is with claims of the form "x causes y." Multiple forms fall within this category, depending upon what kind of terms are substituted for x and y. The most familiar kind of statements has event terms in both the cause and effect place:

(C1)   Franz's outburst caused Sisi's blushing.
 The Battle of Austerlitz caused Austria to sign a treaty with France.
 An asteroid impact caused the mass extinction at the K/T boundary.

Events are particulars—happenings with definite locations and durations in space and time. They involve specific individuals engaging in particular activities and interactions.[5] Accordingly, these are examples of singular causal claims.

---

[5] I will defend this "entity-activity" conception of event and contrast it with more standard views in Chapter 7.

Sometimes singular claims mention objects rather than events as causes:

(C2)   Franz made Sisi blush.
        Napoleon caused Austria to sign a treaty with France.
        An asteroid caused the mass extinction at the K/T boundary.

But these "object-event" claims are just alternate phrasings of event-event claims. It is not Franz or Napoleon or that asteroid that caused an effect, but their acting in some way (yelling, winning a battle, striking the earth). The truth-makers for these kinds of claims will be of the same sort as event-event claims.

In addition to singular causal claims about events, there are general causal claims relating types of events:

(C3)   Exposure to moisture causes wood to rot.
        Concussions cause blurred vision.
        Military defeats cause governments to lose popular support.

Type causal claims are generalizations that range over classes of token causal claims, which are said to instantiate them. For instance, the generalization that concussions cause blurred vision is instantiated by the token causal claim that Franz's concussion caused his vision to blur. Causal generalizations describe causal regularities.

Causal generalizations of this kind may admit of exceptions and still be taken to be true. This is because statements of this kind are implicitly understood either as being hedged by a ceteris paribus clause, or as describing a probabilistic causal relationship. Exposure to moisture generally causes wood to rot, but if the wood is treated with certain chemicals it will not. Similarly, the claim that concussions cause blurred vision is not taken to assert that all token concussions cause blurred vision, but only that they tend to.

Causal generalizations vary widely in scope. They range from very generic claims like "removing oxygen causes fires to stop" to very limited scope generalizations like "questions from Fox News reporters cause the President to perspire."

Many causal claims do not use the word "cause" or its cognates. There are first of all a family of roughly synonymous words and phrases—"produced," "brought about," "changed," "interacted with," "acted upon." More importantly, there are many verbs describing specific forms of activity and interaction, and these activities and interactions are determinate ways of causing. Consider some examples:

(C4)   Sisi tickled Franz.
        The Senate approved the bill.
        The asteroid struck the earth.

These are singular claims about causal processes, though grammatically they are of a quite different form than statements of type C1. Not only do they not contain the word "cause," but there are no terms that refer to events. Nonetheless, of these claims will be corresponding event claims (type C1) that are also true. For instance:

Sisi's tickling of Franz caused Franz's giggling.
The Senate's vote caused the bill's approval.
The asteroid's motion caused its striking of the earth.

While event causal claims are related to the activity claims, the content of activity claims is somewhat different. This is because event causal claims always describe effects, whereas activity claims do not typically do so. For instance, an event causal claim might describe Franz's laughing as the effect of Sisi's tickling, whereas the corresponding activity claim does not describe the activity's effect. The claim that Sisi tickled Franz could be true even if it did not make Franz giggle. All interactions will have some effects—that is part of what it is to be an interaction—but activity claims often do not specify what those effects are.

But although these different claims have somewhat different content, the truth-makers for both kinds of claims will be the occurrence of particular activities and interactions at some particular place and time in the world. It is in this sense that the activity claim is more fundamental. For when Sisi causes Franz to giggle, this causing is something she does by tickling. There will never be causing as such, but only specific activities and interactions that can produce changes.

As with event causal claims, activity claims come in general as well as singular forms. General activity claims express facts about how kinds of activities or interactions produce changes in kinds of entities.

(C5)   Female mammals suckle their infants.
        Water dissolves salt.
        Stars radiate light and subatomic particles.

It is possible, as with singular claims, to recast these claims in event language. For instance, we can say that placing salt in water (one event type) causes the salt to dissolve (another event type). Additionally, the same sort of relation holds between singular and general activity claims as holds between singular and general event causal claims: general activity claims are understood to express tendencies or dispositions that need to be hedged by ceteris paribus classes.

A final class of causal claims that we need to consider are causal relevance claims. Like event causal claims, and activity claims, these claims come in both singular and general forms. Here are some examples of singular causal relevance claims:

(C6)   Franz's weight was causally relevant to his heart attack.
        Obama's race made a difference to his electoral margin.
        The evolution of life on earth depended upon the presence of abundant water.

What distinguishes causal relevance claims from the event causal and activity claims is that they are not about entities, activities, and events, but about *properties* or *features* of those entities, activities, and events. Franz's weight is not an entity, activity, or event, but is a property of Franz (an entity) that makes a difference in the occurrence of Franz's heart attack (an event).

As these three examples indicate, there are a number of English phrases that can be used to express the idea of a property or feature being causally relevant to some occurrence: for instance, an occurrence may be said to depend upon the feature, or the feature may make a difference to or influence the occurrence, or we may say that an event occurs in virtue of the property or feature.

Causal relevance claims are often made about general classes of events, and we can find general claims related to each of the singular claims above:

(C7)   Obesity is causally relevant to heart disease.
       Race makes a difference in elections.
       The evolution of earth-like life forms depends upon the presence of abundant
       water.

As with type-level event claims, type-level causal relevance claims can be understood as representing general tendencies. Race, for instance, makes a difference in many elections, but it need not make a difference in all elections.

Let us summarize the result of our quick survey of types of causal statements. That survey suggests two important distinctions between types of causes, or at least two types of causal claims. On the one hand we have the distinction between singular and general (or token and type) causal claims. On the other, we have a distinction between causal claims about activities and events, which involves entities (parts, objects) engaging in activities and interactions to produce changes, and causal relevance claims that characterize properties that make a difference in the outcome of causal processes. In this and the following chapter we will have to unravel the relationships between these kinds of claims and their truth-makers.

6.3 Singularism and Intrinsicness

While there is a clear distinction between singular and general causal statements, we have not yet considered the ontological grounds upon which the truth of these different kinds of statements depend. We can distinguish two different approaches to causation based upon which of the kinds of relationships described by these statements is more fundamental. Causal generalists hold that singular causal relations obtain because they are instances of a general causal relations (i.e., laws or regularities). Consider the pat of butter I melted while making dinner last night. According to the causal generalist, what makes it true that the heat I applied to the butter last night caused it to melt is that there is a general causal law that whenever butter is heated past a certain point it will melt. The singularist position is the converse. Singular causal relations can obtain even if they are not instances of causal regularities or laws, and what makes causal generalizations true, when they are true, is that they correctly describe a pattern of singular instances of causally related events. The fact that last night my heating the butter caused it to melt is basic, and the general claim that heating butter causes it to melt is true only because it happens that in most or all of the individual cases, heating butter causes it to melt.

Two classic statements of the singularist position come from C. J. Ducasse and Elizabeth Anscombe, both of whom articulated versions of singularism in opposition to then-dominant Humean approaches. Anscombe puts the core singularist intuition this way:

Causality consists in the derivedness of an effect from its causes. This is the core, the common feature, of causality in its various kinds. Effects derive from, arise out of, come of, their causes. For example, everyone will grant that physical parenthood is a causal relation. Here the derivation is material, by fission. Now analysis in terms of necessity or universality does not tell us of this derivedness of the effect; rather it forgets about that.

If A comes from B, this does not imply that every A-like thing comes from some B-like thing or set-up or that every B-like thing or set-up has an A-like thing coming from it; or that given B, A had to come from it, or that given A, there had to be a B for it to come from. Any of these may be true, but if any is, that will be an additional fact, not comprised in A's coming from B.   (Anscombe 1993, 92)

Similarly, Ducasse claims:

[T]he cause of a particular event [is defined] in terms of but a single occurrence of it, and thus in no way involves the supposition that it, or one like it, ever has occurred before or ever will again. The supposition of recurrence is thus wholly irrelevant to the meaning of cause; that supposition is relevant only to the meaning of law. And recurrence becomes related at all to causation only when a law is considered which happens to be a generalization of facts themselves individually causal to begin with.   (Ducasse 1993, 129)

We can probably agree that singularist intuitions like these count among the platitudes of the folk conception of causation, but intuitions are not beyond dispute. These are the very intuitions that Hume sought to undermine with his skeptical argument. Hume's generalist position famously derives from his argument that we cannot directly observe causal connections, and that our concept of cause must arise from our perception of regularities, and much of Anscombe's and especially Ducasse's counterargument proceeds by suggesting that we can in fact observe singular causal relations.

I believe Bogen (2008c) is correct to suggest that the whole question of the observability or non-observability of the causal relation is a red herring. The Humean account of perception and learning is flawed, and it would be a mistake to assume on the basis of Hume's arguments that we cannot at least sometimes observe single-case causal connections. But more importantly, as Bogen points out, the idea that we cannot form concepts that are not directly derived from sense experiences is hopelessly out of date. Causation is a theoretical term, and it is possible for us to understand such terms and have reason to believe that the terms refer, provided the theory is fruitful. None of this is to deny the importance of observations of regularities and various statistical methods to inferring causal connections. The singularist does not deny the epistemic value of regularities, only the claim that causes reduce to regularities.

The distinction between singularist and generalist conceptions of causation is related to another distinction common in the contemporary literature on the metaphysics of causation—namely the distinction between intrinsic and extrinsic conceptions of causation. Menzies argues that intrinsicness is a crucial platitude about causation:

> [T]he causal relation is an intrinsic relation between events. By this I mean roughly that it is a relation between events determined by the intrinsic properties of the events and of what goes on between them. In taking this to be a platitude of the folk theory of causation I am assuming that the folk theory simply contradicts Hume: the commonsense conception presupposes that when events are causally related, there is some connection between them which is determined by the intrinsic nature of the events and of the local spatiotemporal region containing them: the causal relation is not an extrinsic relation depending on a regularity or any other pattern of events happening outside the local spatiotemporal region of the causally related events.   (Menzies 1996)

Insofar as the platitudinous version goes, this is very much the singularist/generalist distinction by another name. If the causal relation is intrinsic, then what makes the relation hold between two events are the intrinsic properties of those two events, and of the local spatiotemporal region (i.e., the process) that connects them. As David Lewis (2004) notes, if this platitude is right, it spells doom for both standard regularity and counterfactual accounts. This is obviously so for the regularity account, since singular causal relations will obtain virtue of instantiating a general pattern or regularity. It is less obviously, but equally true for the counterfactual theory, because, at least on Lewis's version of the counterfactual theory, the truth of the counterfactuals that determine the truth of singular causal claims depends to a significant degree upon what laws obtain in the world in which the claims are being evaluated, and these laws are, at least in part, regularities that are extrinsic to the cause and effect.

By contrast, the ontological account developed in Chapter 2 demands a singularist approach to causation, wherein causation is an intrinsic relation. Mechanisms are particulars—organized collections of entities whose activities and interactions take place at some particular place and time. Laws do not govern these interactions; rather, the laws describe how mechanisms behave, and mechanisms explain laws.[6]

6.4 Production and Relevance

Much of the recent philosophical literature on causation has centered on causal pluralism, and specifically on the idea that there may be two concepts of cause, or two

---

[6] I do not mean here to imply that all laws depend upon mechanisms, but only that most of the laws and generalizations that we use to characterize causal relationships do so. In Section 7.5, I will discuss the question of whether there are fundamental causal laws that do not depend upon mechanisms.

kinds of causal relationships. On the one hand causes are thought to *produce* or bring about their effects. On the other hand, causes are said to be *relevant* to or *make a difference* to their effects. I shall call these different kinds of causal relationships production and relevance. [7]

Productive causal relationships are singular and intrinsic. They involve continuity from cause to effect by means of causal processes. Anscombe's example of parenthood clearly illustrates the productive conception. When a woman gives birth to a child, there is first a productive process whereby the woman's egg is fertilized, and then a set of developmental processes whereby the fetus grows and there is a process of birth by which the child emerges from and is detached from the mother.

The notion of causal relevance on the other hand is essentially comparative. When, for instance, we say that obesity is causally relevant to heart disease, we are drawing the contrast between the case in which obesity is present and the case in which it is not, and saying that the presence of obesity makes a difference. With respect to general causal claims, we can understand difference-making in terms of populations. For instance, within contemporary human populations the frequency of heart disease among the obese is higher than among the non-obese.[8]

While the idea of causal relevance is most easily understood in the case of general causal claims, it can also be applied to singular causal claims. In the single case, the comparison is counterfactual. If we say that Joe's obesity was causally relevant to (or made a difference to) his heart attack, we are implicitly saying, counterfactually, that had Joe not been obese, he would not have had a heart attack, or at least he would have been less likely to have a heart attack.

The distinction between causal production and causal relevance may be easily missed because most events that are involved in causal production also make a difference to what is produced. For instance, attachment of a fertilized egg to the uterine lining both produces and makes a difference to the birth of a baby. But there are some cases where we appear to have production without relevance or relevance without production. For instance, in cases of overdetermination, where more than one cause may be sufficient for an effect, it appears that there is production without relevance. A standard example is the firing squad. Each of the shots fired are sufficient for the prisoner's death, so a single shot is not a difference-maker; but there is a productive

---

[7] The best-known version of this distinction comes from Ned Hall (2004), who refers to the two kinds of causes as production and dependence. Strevens (2008; 2013) makes a similar distinction between causal influence and causal difference-making. The distinction is also closely related to the distinction between intrinsic and extrinsic approaches to causation (Menzies 1996; 1999). My own account of this distinction has been developed in (Glennan 2009; 2010a). I prefer to speak of relevance rather than difference-making or dependence because of the long history of the term "causal relevance" in the literature on causal explanation, and because historically causal-mechanical approaches to causation have been criticized for their failure to capture causal relevance. I will address these issues in the next chapter.

[8] Such correlations are not sufficient for causal relevance. It might be, for instance, that the probabilistic dependence of heart disease on obesity is the product of a common cause.

causal process connecting the gun's firing to the prisoner's dying. Conversely, cases of causation by omission involve difference-makers that are not causally productive. For instance, someone's failure to hit the brake may cause (in the sense of make a difference to the occurrence of) a collision, but there is no productive relation between the omission and the collision.

Another much-discussed case comes from Ned Hall (2004, 236–7). Suppose Suzy and Billy are expert rock throwers, and each throws a rock at a nearby bottle, with Suzy's rock arriving a moment before Billy's, shattering the bottle. Suzy's throw clearly caused the bottle's shattering in the productive sense, as there is a process connecting the throw to the shattering. Nonetheless it does not in this case make a difference to the shattering. Because of Billy's throw, the rock would have shattered either way. Conversely, we can have relevance without production, as in the following case: Suzy throws a rock at the bottle, but Billy throws a rock toward the path of Suzy's rock so that it will knock Suzy's throw of course and prevent the shattering. Meanwhile Tommy throws a rock that intercepts Billy's rock, thus permitting Suzy's rock to strike the bottle. In such a case, Tommy's throw makes a difference to the shattering (and in this sense causes it) even though his throw does not produce the shattering.

While these examples mark an intuitive distinction between the concepts of production and relevance, they cannot by themselves establish that there really are two different sorts of causal relations. Advocates of the primacy of either productivity or relevance can re-interpret these scenarios so as to avoid the pluralist position. For instance, an advocate of a relevance approach can argue that the first scenario is not a case of causation without difference-making: Suzy's throw really did make a difference to the shattering of the bottle, because the shattering caused by Suzy's throw is really a different event than the shattering that would have been caused by Billy's throw. Similarly, the advocate of the production approach can argue that the second scenario is not a case of causation without production, for they can argue that it really is not the case that Tommy's throw caused the shattering, but only that it is explanatorily relevant to the shattering.

While I grant that production and relevance are two different concepts of cause, I will argue that production is fundamental. Both production and relevance claims are made true by objective features of mechanisms that underlie causal relationships, but I will argue that causal relevance claims are ultimately comparative claims about actual or possible productive causal mechanisms. The deeper pluralism of the mechanistic approach is not the pluralism of production and relevance, but the pluralism that suggests that there are many kinds of causal production corresponding to the many ways that different activities and interactions can produce change.[9]

---

[9] On this deeper pluralism, see the discussion of activities in Chapter 2. The varieties (and reality) of causal pluralism has been much debated over the last decade. See, for instance, Campaner and Galavotti (2007); Russo and Williamson (2007); Godfrey-Smith (2009b); Strevens (2013).

In the remainder of this chapter I will help myself to the concepts of production and relevance as I sketch a mechanistic account of the truth-makers of causal claims and compare it to manipulationist and regularity approaches to causation. This will allow us to understand the motivations for and intuitive plausibility of the mechanistic account. The bill will be paid in the following chapter where I spell out in detail a mechanistic account of production and relevance.

6.5 The Mechanistic Theory of Causation

To have an adequate theory of causation is to have an account of what features of the world make causal claims true. But as the examples from Section 6.2 show, there are a variety of kinds of causal claims, and they will have different sorts of truth-makers. In this section I will focus on event claims, first singular (C1) and then general (C3).

Here is a succinct statement of the mechanistic account of the truth conditions for singular event claims:

(MC)   A statement of the form "Event c causes event e" will be true just in case there exists a mechanism by which c contributes to the production of e.

So, for instance, the statement "The heavy rain caused the basement's flooding" will be true just in case there is a mechanism by which the heavy rain contributes to the production of the basement's flooding.

According to this account causal claims are really existential claims about mechanisms (cf. Waskan 2011). In the case of the basement's flooding, what the causal claim asserts is that there is some mechanism (i.e., some causal process) by which some of the rain that fell upon the ground made it into the basement. Perhaps it came from leaking gutters, or perhaps from water flowing toward the foundations because of improper grading. And how and where does this water enter the basement? Perhaps it is seeping through an unsealed concrete block, or maybe it is coming up through a floor drain because of overflowing in the city storm drains to which the floor drain is attached.

Much of the utility of the word "cause" is that it allows us to assert causal dependence without saying anything about the kind of activities and interactions that ground the dependence. Very often we have strong evidence that some event was the cause of another without knowing specifically *how* the cause produced the effect. When I walk into my basement the night after the storm, I see the water and justifiably infer that the storm caused the flood—but I may have no idea just how that water made it into the basement. Causal knowledge of this minimal sort allows us to provide what I call in Chapter 8 bare causal explanations.

It is worth comparing the truth conditions for claims of type C1 with those of activity claims of type C2. One activity claim related to our claim about the flood would be this: "The water from the flood seeped through the cinder blocks and trickled onto the floor." This activity claim is, by contrast to the event-causal claim, a claim about the specific activities and interactions that produced the flooding. It tells us that the way in which the water from the storm produced the flooded basement is by the activities of seeping through the cinder blocks and trickling down the cinder blocks onto the floor. Whereas the event-causal claim asserts the existence of a mechanism, the activity claim tells us something about what this mechanism is.

The truth conditions offered in MC are very weak. Any event that contributes in any way and at any time to the production of some effect will count as causes of that effect. The big bang causes everything, the gravitational field of Pluto contributes to the production of the home run by altering ever so slightly the baseball's trajectory, and so on. Very few of the countless true causal claims are claims worth making, and what counts as a claim worth making will be connected to pragmatic questions about descriptive and explanatory interests.[10]

In keeping with the singularist approach, the New Mechanist account of causation holds that causal generalizations about classes of events (type C3) hold true in virtue of causal relations between their singular instances (Glennan 2011). So, for instance, the claim that heavy rains cause basements to flood will hold true in virtue of the fact that specific heavy rains do (or perhaps would) cause specific floods. But here immediately we see that type causal claims are ambiguous. It is unclear just how many token causal claims must be true for one to appropriately assert the type causal claim. At the one extreme we might offer these truth conditions for general causal claims:

> A statement of the form "Event type C causes event type E" is true just in case there exist event tokens c and e of types C and E, where there exists a mechanism by which c contributes to the production of e.

If this is all that is required for type causal claims to be true, then we will have a lot of true claims. For instance, it will be true that diving into swimming pools causes heart attacks so long as there is one instance in which there was a mechanism by which diving into a pool contributed to the production of the heart attacks.

Presumably causal generalizations are meant (at least on most occasions) to assert more than this. Here are some possible alternatives:

> A statement of the form "Event type C causes event type E" is true just in case there are multiple event tokens c and e of types C and E, where in each case there exists a mechanism by which c contributes to the production of e.

> A statement of the form "Event type C causes event type E" is true just in case, more often than not, when there are events c of type C, there are mechanisms by which c contributes to the production of an event e of type E.

---

[10] Philosophers have long recognized that the causes of any particular event are many, and that pragmatic considerations are central to understanding why some of these events are elevated to the status of true causes or even *the* cause of some event. Opinion is divided about whether there is any objective distinction to be had between causes and background or enabling conditions. Classic sources on this issue are Mackie (1965) and van Fraassen (1980). See Strevens (2008) for a more recent take on the issue.

A statement of the form "Event type C causes event type E" is true just in case, whenever there is a token c of type C, there is a mechanism by which c contributes to the production of an event e of type E.

These four interpretations of general causal claims all make the truth of general causal claims dependent upon the truth of singular causal claims, but they differ in the number of singular instances required: at least one instance, more than one instance, the majority of instances, or all instances. We could stipulate that one of these (or some other) conditions are what is required for a causal generalization to be true, but such a stipulation is not required or helpful. General causal claims have no single analysis, and the proper use of such claims will be regulated by the purposes and interests of individuals and communities using such claims. Take for instance the claim that heavy rains cause basements to flood. I would say that this claim is true, and true because flooded basements are a fairly common occurrence. It does not happen to everybody every day or every year, but it has happened to me and my friends, on more occasions that I can count. It happens enough to matter—enough to evaluate risks and seek remediation.

Other difficulties arise because of the vagueness of event types. As we have already observed, judging whether tokens are members of kinds (kinds of entities, activities, events, mechanisms) are judgments of similarities, and the degree and respect of similarity required to judge two events of the same kind will depend upon the use to which we are putting such judgments. What is enough to make a rain heavy, for instance?

While type causal claims assert a general mechanistically mediated relationship between causes and effects, they do not require that the same kind of mechanism will account for the causal relationship in each case. Sometimes the mechanism will be the same: strychnine kills in the same way every time; but often the mechanisms are various. Take the generalization that alcoholism causes ruin. Certainly it can, but the roads to ruin are various: it can destroy livers, bank accounts, and relationships.

Given the various ambiguities, we will have to live without a single simple account of the truth-makers for causal generalizations and content ourselves with the basic singularist insight that causal generalizations are made true by their instances. But precisely because of the fact that the truth of general causal claims is derivative on singular causal claims, we can live (for purposes of explicating the nature of causation at least) with a focus on the truth-makers for singular causal claims.

In the remainder of this chapter I want to flesh out the mechanistic account by means of some more detailed and scientifically nuanced examples. These examples will clarify the relation between singular and general causal claims as well as between causation and manipulation. The cases we will discuss are connected to the deadly condition that has historically been called puerperal or childbed fever, and is now understood to be one instance of the syndrome generally known as sepsis.

Let us begin our story with the death of Jane Seymour, the third wife of Henry VIII, who died twelve days after the birth of her son Edward. It is difficult to know what caused her death, both because of secrecy surrounding the details of the birth and because no observations or tests were performed that could isolate what modern medical science would think of as the causes of her death. It is, however, clear from reports that Jane showed symptoms of massive infection, and it has been conjectured that she died from puerperal fever. Let us suppose for the sake of argument that this is so. What then caused Lady Jane's death, and what makes the associated causal claim true according to MC? Puerperal fever is no longer a diagnostic category, but it is understood to arise from a bacterial or other infection introduced into a woman in her reproductive tract at the time of childbirth. But puerperal fever is not one kind of thing, and the particular instance of puerperal fever and how it plays out will depend upon particular events. Since we do not know exactly what happened, let us just imagine what might have happened to Jane. Let us suppose that Jane had a midwife (call her Mary), and that Mary introduced the infection into Jane's body when she reached into the birth canal to perform a cervical examination. Call this event c, and suppose that it is true c caused Jane's death (event e).

According to the mechanical theory, this causal claim is true if, and only if, there exists a mechanism by which c contributed to the production of e. The mechanism in this instance is a processual one, in which a collection of entities engages in a sequence of temporally extended activities and interactions that produce, in the end, the outcome, Jane's death. The mechanism works something like this. The surface of Mary's hand touches some surface inside Jane's body—say the cervix. The surface of Mary's hand has on it a population of some bacterium. For definiteness, let us imagine it is a population of *staphylococcus aureus*. The specific surface that Mary touches is lacerated or compromised in such a way that some of this population migrates into Jane's bloodstream. Over time the bacterial cells multiply via mitosis. Their presence triggers an immune response, and it is the excessive immune response, rather than the bacteria directly, which ultimately produce the conditions of septic shock, which include sharp drops in blood pressure, and, in Jane's case, organ failure and death. Regarding sepsis, O'Brien et al. write:

Because sepsis is defined as a syndrome, it is likely that heterogeneous pathophysiologic processes are contained under this single term. The interaction of microbiological products with a host that is susceptible due to genetic or other factors induces a cascade of immunomodulatory mediators, leading to cellular and organ dysfunction. The major pathways involved in sepsis include the innate immune response, inflammatory cascades, procoagulant and antifibrinolytic pathways, alterations in cellular metabolism and signaling, and acquired immune dysfunction.   (O'Brien Jr et al. 2007, 1013)

Notice that sepsis is a term applied to a set of broadly similar but nonetheless heterogeneous pathological physiological processes. There is no one such thing as sepsis. There is, however, Jane's particular instance of sepsis—an instance that involves a particular infectious agent, a particular immune response, a particular set of inflammatory cascades, a particular set of resulting organ failures, a particular death.

So according to MC, the reason that it is true that Mary's touching Jane (*c*) caused Jane's death (*e*) is the fact that there exists this particular mechanism on this particular occasion by which *e* contributes to the production of *c*. It is not hard to see how this particular mechanism exhibits the features of minimal mechanism. There are entities at a number of levels or organization—tissues and organs, cells (e.g., bacteria, macrophages, and lymphocytes) and subcellular molecular components (e.g., Toll-like receptors and cytokines). They engage in particular activities and interactions. For instance, the bacteria divide. The Toll-like receptors bind to molecules from the bacterial cells, and so on. These activities and interactions are organized into various pathways, and it is the collective action of these many agents (on the order of billions or more) that are productive of system-level phenomena like drops in blood pressure, organ failure, and death.

Though this event is a cause of Jane's death, there are countless other events that count as causes by MC, since many events contribute ultimately to the production of this or any event. Here are a few others: Mary had bacteria on the surface of her hands, and they had to come from somewhere. Whatever event or events got them there certainly count as causes of Jane's death; so too do the events that led to the lacerations in the birth canal that permitted the entrance of the bacteria into the bloodstream. Likely these came from Edward's head putting excess pressure on stretched tissues within the birth canal, so this would count as a cause of Jane's death. Henry VIII is responsible too, for Henry's having sex with Jane (on some particular occasion) contributed to the production of Edward, who contributed to the production of the lacerations in the surface that interacted with the midwife's hand. We might even add the many occasions in which Jane ate during her pregnancy, for these events, albeit in a somewhat less specific way than Henry's impregnating Jane, contributed to the production of Edward. We could go on indefinitely, since there are indefinitely many things going back arbitrarily far into the past that are connected via a productive mechanism to Jane's death. While for a variety of pragmatic or epistemic reasons we might identify only certain of these events as the cause or as major causes, such distinctions are not part of MC.

While MC is quite permissive in what counts as a cause of an event, it does insist that the cause is a genuine event. Henry might be tempted to blame his first wife, Catherine of Aragon, for her failure to produce a male heir, and indeed, had Catherine given birth to a son, Jane would not have had the death she had. But, as we will discuss at length in the next chapter, omissions are not genuine events, and while an omission can be causally relevant, it cannot contribute to the production of anything.

6.6 Mechanisms and Generalizations

This very particular story of the causes of Jane Seymour's death can be contrasted with more general claims about childbed fever as a cause of death in childbirth. The disease and its treatment has an important place in the annals of medicine, and Carl Hempel, in his textbook *Philosophy of Natural Science*, uses the case of the Hungarian physician Ignaz Semmelweis's investigations of the causes of childbed fever as a case study illustrating the hypothetico-deductive method of scientific inference. Semmelweis's investigation began with an observation about the difference in mortality rate between two divisions in the Vienna General Hospital. Hempel describes Semmelweis's findings:

In 1844, as many as 260 out of 3,157 mothers in the First Division, or 8.2 per cent, died of the disease; for 1845, the death rate was 6.8 percent, and for 1846, it was 11.4 per cent. These figures were all the more alarming because in the adjacent Second Maternity Division of the same hospital, which accommodated almost as many women as the First, the death toll from childbed fever was much lower: 2.3, 2.0, and 2.7 per cent for the same years.   (Hempel 1965, 3)

Semmelweis explored hypotheses about differences between the two divisions that might explain this discrepancy, ranging from a hypothesis about different birth positions to one about psychological differences induced by the behavior of priests administering last rights. In several cases he performed experiments, by changing a procedure in the First Division to match that of the Second Division, but none of these interventions resulted in changes in the mortality rate. According to Hempel, the causal agent was suggested to Semmelweis by an accident in the First Division. A colleague who had received a cut with a scalpel used during a dissection developed symptoms very similar to those of childbed fever and ultimately died. Semmelweis surmised that material from a cadaver was transferred into his colleague's bloodstream. Recognizing that medical students in the First Division examined women directly after coming from the autopsy room, he ordered interns who had performed autopsies to wash their hands in a solution of chlorinated lime. After this intervention, the mortality rate within the ward dropped dramatically, to a rate less than that in the Second Division. Semmelweis concluded that "cadaveric matter" was the cause of childbed fever. Semmelweis drew this conclusion before the development of the germ theory of disease, but he was in essence correct. Childbed fever is often caused by the introduction of bacteria or other microbes into the bodies of women via unsterilized hands or instruments.

Semmelweis's investigations were not focused on any single case of childbed fever, but sought to understand and mitigate the general causes of the disease within the hospital. Based upon his observations, Semmelweis might likely draw these conclusions:

1. Admission to the First Division causes childbed fever.
2. Examination by interns in the First Division who have recently interacted with cadavers causes childbed fever.
3. Hand-washing by dirty-handed interns in the First Division prevents childbed fever.

These are examples of type-level event claims (type C3). On the singularist approach that I am advocating these are best understood as somewhat vague summaries of token causal claims. But let us consider for a moment how an advocate of a generalist approach to causation might interpret these claims.

If one accepts a generalist approach to causation, what makes singular causal claims true is that they are instances of a general pattern. Psillos describes a simple regularity account of causation that will allow us to explore this possibility. On that account we have the following truth conditions for a singular causal claim:

c causes e iff

i.   c is spatiotemporally contiguous to e;

ii. e succeeds c in time; and

iii. all events of type C (i.e., events that are like c) are regularly followed by (or are constantly conjoined with) events of type E (i.e., events like e). (Psillos 2009, 131)

It is evident that, were this account accepted, none of the three causal generalizations I have listed would ground singular causal claims, since the associations between events are not deterministic, but at best change the probability of events. The only way to make a strict regularity theory work is to argue that the regularities that ground singular causal connections are not of the sort described by these generalizations, but are rather much more detailed and fine-grained. So, for instance, the reason that in one instance a woman examined by a dirty-handed intern contracts childbed fever while another does not has to do with fine-grained physiological differences between the women and their circumstances, which are not captured in the course-grained generalization.

This defense of a regularity approach is very much the position advocated by Davidson (1967; 1995), who suggests that for a singular causal statement to be true, there must be an exceptionless causal law that covers it, even if in most cases we will have no idea what that cause is. Even if you think there are good reasons for adopting Davidson's position (and I do not), the causal laws required by Davidson will have nothing to do with the sorts of causal generalizations that Semmelweis would have made, and that scientists, physicians, and ordinary folk normally traffic in.

Alternatively, one might adopt a probabilistic version of a regularity theory, replacing clause iii of Psillos's definition with a claim to the effect that events of type C are positively correlated with events of type E. On such an account these causal generalizations will be true, and we might infer the truth of singular causal claims from them. For instance, Clara, a woman giving birth in the First Division, is examined by an intern, Franz (with unwashed hands), and develops puerperal fever. We conclude that Franz's examination caused Clara's puerperal fever, because it was spatiotemporally contiguous to the fever, preceded it in time, and fell under the probabilistic generalization regarding the relation between examinations and childbed fever.

The inference from the causal generalization to the singular causal claim is unproblematic; we make such inferences all the time. However, it does not follow from this that we have the truth conditions for these causal claims right. On a regularity theory (probabilistic or otherwise), what makes the singular causal claim true is that it is an instance of the general causal claim. But this seems backward. The general causal claim seems simply to be a summary claim about singular causal events. The general claim that dirty-handed examinations probabilistically cause childbed fever follows from the fact that in some instances particular dirty-handed examinations produce childbed fever, not the other way around.[11]

There are a couple of reasons to prefer the singularist interpretation. In the first instance, it is puzzling in general that the truth of a causal claim about a single instance (e.g., Clara's fever) should depend upon the truth of claims about other instances. Also, in the probabilistic case, we should acknowledge that even if it is the case that a certain type of event raises the probability of another type of event, it might not be the case in the particular instance that the probability-raising event was actually causally implicated. For instance, Franz might have examined Clara with dirty hands without actually having introduced any bacteria into her bloodstream. Clara might subsequently develop puerperal fever on the basis of an infectious agent introduced by some other event. Finally, we should note that the sort of causal generalizations appealed to here are actually of quite restrictive scope. They concern a certain population of women at a particular place and time. And while disinfecting one's hands is no doubt good practice in hospitals everywhere, the degree to which any particular hand-disinfecting makes a difference will depend upon the particular context of the individual case.

At root, the difficulty in the application of the regularity approach lies with its appeal to event types. As we have suggested above, the event types "dirty-handed examination" and "puerperal fever" do not represent uniform kinds of things, but in fact are ways of classifying similar but not identical groups of individual events. Events and the mechanisms that connect them are always and everywhere particular.

6.7 Mechanisms and Manipulations

The puerperal fever examples have already lent credence to a mechanistic account of causation and offered something by way of explanation of what is meant by saying that there exists a mechanism by which c contributes to the production of e. In this section, I would like to offer some further elucidation by contrasting the mechanistic account with Jim Woodward's well-known manipulationist/interventionist account. We will be aided in this by a pair of papers (Waskan 2011; Woodward 2011) in which Woodward and Jonathon Waskan offer competing accounts of the relationships between mechanisms and manipulations.

Woodward understands his account of causation to fall under the aegis of difference-making (DM) approaches to causality. According to Woodward,

DM accounts share several common commitments. One is that causal claims involve a *comparison* of some kind between what happens in a situation in which a cause is present and

---

[11] For more on the mechanistic account of probabilistic causal claims see Glennan (1997a; 1997b) as well as the discussion of stochastic mechanisms in Chapter 5.

alternative situations (which may be actual or merely possible) in which the cause is absent or different.   (Woodward 2011, 411)

The contrast need not be between actual situations, but may also be between the actual situation and a possible but non-actual situation. For all DM accounts, "the claim that C causes E is taken to 'point beyond' the local, actual situation in which C and E occur and has implications for what happens or would happen in alternative situations in which C does not occur" (411).

Woodward contrasts DM accounts with geometrical/mechanical (GM) accounts, which he sees as claiming that

whether there is a causal relationship between two events c and e just has to do with whether c and e occur and whether there is an appropriate "connecting" process (or mechanism) between them; moreover whether there is such a process depends just on what is actually true of the particular occasion of interest, and does not depend on what does or would happen on other occasions.   (413)

The distinction between DM and GM accounts parallels the distinctions we have made between relevance (DM) and production (GM), and between extrinsic and intrinsic accounts of causation. But Woodward is no causal pluralist, because he believes DM accounts are foundational.

At the same time, Woodward does not see the difference-making approach as hostile to the idea that understanding mechanisms is central to much of the scientific enterprise. He remarks that it is "common ground" that "the identification of mechanisms is a major goal of theory construction, information about mechanisms is important in causal explanation and so on" (409–10). But Woodward believes that ultimately information about mechanisms reduces to information about difference-making. Woodward cites as an example the causal generalization that aspirin causes pain relief. This basic causal claim can be established on interventionist grounds, but we also know something of the mechanism by which it produces this effect. Woodward remarks that:

On one natural way of understanding this mechanism information, it seems to fit well into a difference-making (and interventionist) framework—what is provided includes more detailed, fine grained difference-making information (and relatedly, more detailed, fine-grained possibilities for intervention) that goes well beyond information about the overall difference-making relation between aspirin and pain relief.   (419)

In Woodward's view mechanisms just provide "more of the same"—information about fine-grained causal relations interpretable in interventionist terms.

This is not the only way to interpret the relationship between manipulations and mechanisms. The alternative is to see mechanisms—actual this-world mechanisms—as the truth-makers for causal claims (Waskan 2011). It is because such mechanisms exist that Woodwardian interventions have the effects that they do. This alternative approach is in keeping with Canberra planning. The manipulability account is a piece of conceptual analysis (and a highly empirically informed one) that tells us what causes do: causes are the things that allow us to manipulate and control the world. But this analysis does not yet tell us what causes are—and that is where we need mechanisms. Mechanisms provide the ontological grounding that allows causes to make a difference. This is the approach I wish to advocate. While in his most extended discussions of mechanisms (2000; 2011), Woodward clearly rejects the Canberra approach, it is consistent with much that Woodward himself has to say. Woodward characterizes his project as a semantic or interpretive one rather than as a metaphysical one (2003, 38, 95). Additionally, his account is explicitly non-reductive, defining the concept of cause by reference to the concept of intervention, where intervening is itself a special kind of causing. He is concerned principally with characterizing the conceptual relations between causation, manipulation, evidence, and explanation.

Woodward's analysis of causation in terms of manipulation admits of two readings. First, there is a semantic/epistemological interpretation according to which the manipulability account provides a modified Canberra-style analysis but which is open on what in nature ontologically ground manipulations. Alternatively, there is an ontological reading according to which difference-making accounts, and in particular the manipulationist view, provide an account of the basic truth-makers for causal claims. Causal claims are true, just because certain counterfactual relationships hold. On this view, mechanism talk is just a way of talking about fine-grained difference-making relationships. Let us call these two readings the epistemological Woodward and the ontological Woodward.

While Craver (2007) has raised doubts about intrinsic/mechanistic accounts of causation (Glennan 1996; Salmon 1984; Dowe 2000) and has utilized Woodward's manipulability account within his treatment of mechanistic explanation, Craver does not believe Woodward has specified a clear and viable ontological view. Craver writes:

One can complain that the manipulationist account presupposes a metaphysics of causation, and refuse assent until an account of the metaphysics is provided, or one can recognize the manipulationist account of causal relevance as a normative framework that any adequate metaphysics should satisfy, or better, explain. I do not discuss here whether such metaphysics is required or what the available metaphysical options are. Even if the manipulationist view does not identify the truth-maker for causal claims, it is nonetheless an illuminating analysis of the causal truths themselves, and it is crucial for the project of deciding which putative metaphysical explanations (that is, which truth-makers) are adequate and which are not.   (Craver 2007, 105–6)

I can concur here with Craver, and his remarks are completely in keeping with the Canberra approach.[12] The manipulationist approach characterizes central aspects of

---

[12] Some have thought my own decision to characterize causal interactions (Glennan 2002a) using Woodward's language of invariant generalizations was acceding to the view that causal interactions must ultimately be understood in interventionist terms. While understandable, this is actually a misinterpretation of my intended view. While I said that interactions could be characterized in terms of Woodward's invariant change-relating generalizations, I did not say that causal relations were defined in terms of them.

the role that causes play, and so imposes a constraint on metaphysical accounts. A metaphysical account should "satisfy, or better, explain" the manipulationist account of causal relevance.

While Craver's remarks suggest he has adopted the epistemological rather than the ontological reading of Woodward, Woodward himself seems pushed to the ontological reading. He is pushed, I think, because he believes that no extent non-comparative intrinsic account of causal production can play the ontological role needed. Geometrical/mechanical accounts fail to explain the relation of causal relevance, and they also have difficulties handling cases of negative causation—e.g., causation by omission and disconnection. The objections that Woodward and Craver raise to such accounts are serious ones, which I shall address in the next chapter. For the moment, however, I want to press a negative case against the ontological Woodward, contrasting it with what Waskan (2011) calls the "actualist mechanist account" and arguing that difference-making does not give us a plausible account of the nature of the causal relation.

So let us take seriously an ontological view of Woodward—one in which counterfactuals about the results of ideal interventions are the truth-makers for causal claims. What does this view look like? Interestingly, while Woodward's theory of causal explanation is a type-level theory, his views on the truth-makers of causal claims appear to be singularist. He writes:

Although there is a distinction between type and token-causal claims, it does not follow that there are two kinds of causation—type and token—or that in addition to token-causal relationships involving particular values of variables possessed by particular individuals, there is a distinct variety of causal connection between properties or variables that is independent of any facts about token-causal relationships. In my view, a claim such as "X is causally relevant to Y" is a claim to the effect that changing the value of X instantiated in particular, spatiotemporally located individuals will change the value of Y located in particular individuals. Thus, the truth of a claim such as (S) "Smoking causes lung cancer" depends on relationships that do or would obtain (under appropriate manipulations) at the level of particular individuals... (Woodward 2003, 40)

Thus, on Woodward's view, we may offer type-level causal explanations of childbed fever, which cite dirty-handed examinations as a cause, but the truth of such type causal claims will depend ultimately upon the fact that certain singular causal relations obtain between particular examinations and particular fevers. For this reason Woodward's account of causal relations is a singularist one.

To see the difference between the mechanical and manipulationist approach to causation, we must focus on the truth conditions for single-case causal claims. Roughly put, Woodward holds that a singular causal claim of the form *c* causes *e* will be true if

---

In particular, I suggested that facts about mechanisms explain why these generalizations are true. Equally important, while Woodward closely ties together intervention and invariance in his (2003), this connection is not essential. It is possible to understand the notion of invariance without adopting a manipulationist view of causality.

intervening on c would change the probability of e. For instance, it is true that Mary's dirty-handed examination caused [and let us for the sake of argument suppose this is a deterministic case] Jane's death because if one were to manipulate Mary (by say washing her hands or stopping the examination), then Jane would not die that death. In contrast, on the actualist-mechanist account, Mary's dirty-handed examination caused Jane's death because there was an actual mechanism by which the examination actually contributed to the death. What is the difference between these accounts, and what is there to say in favor of each?

The most obvious difference here is that on the manipulationist account, the truth of the causal claim depends upon the truth of a counterfactual, whereas on the mechanist account the truth depends upon the existence of an actual mechanism. On Woodward's account, the counterfactual in question is what Woodward calls a "same object counterfactual." What makes this claim true is a claim about what would happen if you intervened on Mary's particular dirty-handed examination, not on dirty-handed examinations in general. This again is Woodward's singularism.

But we cannot get a clearer handle on just what, according to Woodward, makes causal statements true, unless we can say more about the truth conditions for the singular counterfactual claim. Woodward does not appeal explicitly to any particular account of the meaning of counterfactuals, but given his insistence that difference-making is fundamental, it seems we must interpret the counterfactuals literally as claims contrasting the actual world with possible worlds in which an intervention occurs. For all DM accounts, "the claim that C causes E is taken to 'point beyond' the local, actual situation in which C and E occur and has implications for what happens or would happen in alternative situations in which C does not occur" (Woodward 2011, 411).

In fact, if we combine these remarks about the comparative character of DM accounts with Woodward's singularist view that type-level causal claims are made true by singular causal claims holding between individuals, we are led to the conclusion that the truth-maker for a singular causal claim is a claim about the contrast between the actual situation in which a cause leads to an effect and a non-actual but possible situation in which an intervention on the cause alters this. That is, we must take literally the view that the truth-maker for the causal claim is a claim about what is the case in certain non-actual but possible worlds—which entails accepting some form of realism about possible worlds.

There is an alternative that I prefer, which suggests that counterfactual locutions—while they may perhaps be usefully analyzed in terms of the semantic framework of possible worlds—do not literally make claims about existent (but non-actual) possible worlds. They instead provide a way of characterizing certain modal facts about the actual world.[13] In particular, counterfactuals are convenient (perhaps even ineliminable)

---

[13] On some accounts, the notion of a modal fact about the actual world is incoherent. Modal facts inevitably, on this view, say something about non-actual worlds—but this begs the question against views that relations of determination or production can be understood as features of the actual world. I shall return to this subject in Chapter 8.

ways of expressing relations of causal determination that are intrinsic actual-world relations rather than contrastive relations between this world and other worlds. Adopting this interpretation of counterfactuals decreases the distance between the mechanistic and the manipulationist position.

But even if one adopts such an account of the meaning of the contrastive counter-factuals, there remains an essential point of difference between the mechanist and the manipulationist about which are the more basic truth-makers for singular causal claims. Both mechanists and manipulationists hold that there are such things as mechanisms and that manipulability relations are central to our understanding of causation, but, for the mechanists, manipulability relations are explained by mechanisms, while for the manipulationist, mechanistic information is simply finer-grained information about difference-making. Waskan puts the mechanist's concern with Woodward's position nicely: "The driving intuition behind some mechanists' resistance to Woodward's account of the contents of causal claims is that causal claims make assertions about what *actually* happens rather than about what would happen...In particular, some contend that causal claims make assertions about actual productive *mechanisms* connecting cause and effect" (Waskan 2011, 394).[14] Jane Seymour's death illustrates the point. The content of the claim that Mary's examination caused Jane's death is that there existed an actual mechanism by which the examination contributed to the production of the death. Because there was such a mechanism, it follows that certain interventions (e.g., Mary washing her hands) might have prevented the death. But it is the facts about the actual mechanism that explain the truth of the claims about possible interventions rather than the other way around.

6.8   Conclusion

My aim in this chapter has been to explore the relationship between mechanisms and causation, and to situate a mechanistic approach to causation within the conceptual frameworks and debates familiar from the philosophical literature. We have seen that the mechanistic approach is a singularist one, that it supposes that cause and effect are intrinsically related via mechanical processes, and that it is the existence of these processes upon which the truth of a variety of causal claims, both singular and general, depend. We have also explored the relationship between mechanisms, manipulations, and counterfactuals, and I have argued, contra James Woodward, that mechanisms come first, in the sense that the causal efficacy of manipulations comes from mechanisms, and that the truth-makers for counterfactual claims about possible interventions are actual mechanisms.

---

[14] Waskan emphasizes (as I have in Section 5.5) that the mechanist cannot reasonably be committed to the view that causal claims make claims about specific causal mechanisms, for it is clear that we often make causal claims in cases where we are unaware of what sort of mechanism may be involved. Rather, the mechanist holds that the causal claim amounts to the claim that *there exists* some actual productive mechanism, even if the nature of the mechanism is unknown.

This account of the actualist mechanist approach should, I hope, have made clear the intuitive plausibility of the position. But more needs to be said. While we have said what actual production is not, we have not really spelled out what it is. In the next chapter I will set out a new account of mechanism-dependent production that will remedy this defect. This account will allow us to respond to a number of criticisms that have been made of geometrical/mechanical approaches to causation. Using this account of mechanism-dependent production, I will be able to show how a New Mechanist theory of causation can avoid certain difficulties that have plagued other geometrical/mechanical approaches to causation.

7

Production and Relevance

In Chapter 6, I introduced the idea that there may be two distinct kinds of causal relationships. Causal productivity involves transmission of something from cause to effect via a causal process. Causal relevance describes a relationship wherein a cause makes a difference to an effect. But so far we have said little, beyond appealing to intuitions and examples, to sort out just what the relation of production is, and how it is related to relevance and difference-making. My aim in this chapter will be to remedy this defect by offering a mechanistic account of production and relevance.

The idea of production is central to all of the family of theories that Woodward (Section 6.7) classifies as geometrical/mechanical. But the central argument of this chapter will be that the most familiar accounts of causal production have got it wrong, and that the New Mechanism can provide a more adequate account of causal production. The crux of the difficulty with these theories is that they assume that a proper theory of causal production will be a theory of what Dowe has called "physical causation." That is, they have assumed that the place to look for an account of causal production is at the level of fundamental physics. Take, for instance, Michael Strevens (who uses the term "influence" for what I am calling "production"):

We humans are disposed to read fundamental physics, I think, as describing a web of causal influence in which many fundamental-level facts come together to causally bring about, by way of the fundamental laws, other fundamental facts.   (Strevens 2013, 306)

Strevens's idea is that this set of fundamental web of influence is what ultimately grounds all of our claims about causes. Causal explanation for Strevens is a matter of difference-making, but those differences are ultimately grounded in features of this fundamental web of influence.

Some doubt that this web of fundamental causal influence should really be called causal at all—so they give it a different name. For reasons obscure, a number of philosophers have taken to calling it biff. David Lewis, responding to Menzies's arguments that causation must be an intrinsic relation, describes the relation between causation and biff in this way:

What is causation? As a matter of analytic necessity, across all possible worlds... [i]t is somehow a matter of counterfactual dependence of events (or absences) on other events (or absences).

What is causation? As a matter of contingent fact, what is the feature of this world, and of other possible worlds sufficiently like it, on which the truth values of causal ascriptions supervene?— It is biff… Biff is literally the basic kind of causation, in this world anyway: the basis on which other varieties of causation supervene.   (Lewis 2004, 287)

Lewis thinks that biff is not really causation, because he thinks (using a Canberra-style analysis) that the biff role and the causation role are different. Nonetheless, he grants that biff is in a certain sense a causal relation. As a matter of contingent fact there must be some web of influence (the biff relations) upon which all of this difference-making causation depends. The fact that Lewis talks about biff as a "basic kind of causation" upon which everything else supervenes suggests that he too accepts that biff (or production or influence) is a matter of basic physical processes.[1]

Contrary to Lewis, and in agreement with Menzies, Dowe, Strevens, and others, I think that production (or influence or biff) really is causation, in the sense that I think that some idea of production is really part of our concept of cause. But what relations get the honorific "causal" is really not the most important issue here. What matters is that there are surprising levels of agreement, even among defenders of extrinsic conceptions of causality, like David Lewis, that relations of causal difference-making depend ultimately on patterns of production. And with this I agree.

Where this analysis goes wrong is in assuming that the only proper place to look for production is in physics. We can and do apply the concept of production to relations between entities at many levels "above" that of physics, and we need not adopt a stance that all that production talk is unreal. Yes, a gravitational field may produce a change in the trajectory of a particle, but so too may sex produce children, and eating produce a supply of nutrients and waste products. As we discussed in Chapter 2, the thing that unifies the diverse panoply of things we call activities is that all activities produce change.

The mechanistic account of production that I will develop below allows for genuine productivity in activities and interactions, but before turning to this theory it would be useful by way of contrast to spell out in a bit more detail what a physical theory of production might look like. To that we now turn.

7.1 Physical Theories of Causal Production

Within the causation literature of the last fifty years, the most prominent early advocate of a production-based approach to causation was Wesley Salmon. Salmon's turn to the production approach in the late 1970s and early 1980s arose from a growing dissatisfaction with logical empiricist approaches to scientific explanation. According to Salmon's deeply influential account of the development theories of scientific explanation (Salmon 1989), the most prominent models of scientific explanation, including

---

[1] For a similar interpretation about the need for some biff, see Handfield et al. (2008).

the deductive nomological, inductive statistical, and statistical relevance approaches were all examples of "inferential approaches." Explanations were arguments, and to explain something was to show how some statement describing the explanandum could be inferred from a set of other statements (the explanans). Salmon came to believe that the problems with the inferential conception were intractable, and fundamentally misconstrued what explanations were. In its stead, Salmon proposed an ontic conception of explanation, where explanations are explanatory in virtue of fitting the phenomenon to be explained into the "causal structure of the world." What is the causal structure of the world? According to Salmon, it is the set of causal processes and interactions through which causal influence is propagated.

According to Salmon, the key to understanding causal relations was to explicate two notions, which he called production and propagation. In his initial introduction of these concepts he does not offer a definition, but instead offers examples "familiar to common sense":

When we say that the blow of a hammer drives a nail, we mean that the impact **produces** penetration of the nail into the wood. When we say that a horse pulls a cart, we mean that the force exerted by the horse **produces** the motion of the cart. When we say that lightning starts a forest fire we mean that the electrical discharge **produces** ignition. When we say that a person's embarrassment was due to a thoughtless remark we mean that an inappropriate comment **produced** psychological discomfort.   (Salmon 1981, 49–50).

Regarding propagation (or transmission) he offers these examples:

Experiences which we had earlier in our lives affect our current behavior. By means of memory, the influence of these past events is **transmitted** to the present. A sonic boom makes us aware of the passage of a jet airplane overhead; a disturbance in the air is propagated from the upper atmosphere to our location on the ground. Signals **transmitted** from a broadcasting station are received by the radio in our home. News or music reaches us because electromagnetic waves are **propagated** from the transmitter to the receiver.   (50)

Production, it can be seen from these examples, involves interactions that bring about changes to the things interacting. Propagation involves the transmission of effects of these interactions through space and time; it is propagation that allows the results of interactions to persist.

It is notable that Salmon's examples of production are not drawn from physical theory, but instead appeal to common-sense macro-physical and even psycho-social interactions. Production in this sense involves activities and interactions in the sense discussed in Chapter 2. Similarly, while some examples (signal transmission) involve processes described by physics, others (persistence of memory) do not.

For Salmon, the causal structure of the world is the totality of causal processes that transmit causal influence, together with the totality of causal interactions, which are intersections between these causal processes. Any object in the world is a causal process in the sense that it is something that can transmit influence to other objects in the world. For instance, a baseball and a bat traveling through space are two causal processes that causally interact when they strike each other, producing changes in the trajectories of the baseball and, to some extent, the bat.

Salmon himself identified his approach as "mechanical philosophy" (Salmon 1984, ch. 9), and his account of the causal nexus is related closely to our conception of minimal mechanism.[2] Entities are processes that propagate influence, and interactions are events where these causal processes intersect and modify each other.

But while the guiding intuitions are similar, there remain significant differences between Salmon's mechanical philosophy and the New Mechanism. One difference is that Salmon's account is primarily concerned with the horizontal dimension of mechanical processes, or as it is sometimes called, the etiological aspect. The New Mechanist account grants the importance of the horizontal dimension, but also emphasizes the vertical or constitutive aspect of mechanisms, where the interaction of a mechanism's parts is responsible for the behavior of the mechanism as a whole (see Section 2.2). The other difference is the causal pluralism associated with the New Mechanist's account of activities. Causal production is, for the New Mechanist, not one thing, but many. Salmon on the other hand hoped for a reductive account of the one thing that is causation.

To better understand the difference between Salmon's mechanical philosophy and the New Mechanism, we must attend to a few details of Salmon's account of production and transmission. Famously, he gave two versions: the mark theory and the conserved quantity theory. According to the mark theory (Salmon 1984), genuine causal connections must be mediated by causal processes, and not all processes are causal. Causal processes are distinguished from pseudo-processes on the grounds that causal processes are capable of transmitting marks. For instance, a beam of light is a causal process, because I can mark it (say by putting a filter over the beam) and the mark will be transmitted by the process (as can be seen, for instance, when the light appears colored where it strikes a wall). A shadow, on the other hand is a pseudo-process. If I introduce a mark into a shadow along a wall, the mark will not be transmitted as the shadow moves. Salmon's theory of production was based upon the concept of an interactive fork. An interactive fork is an intersection between two causal processes in which each process modifies (i.e., marks) the other. The mark theory had what to Salmon was the regrettable feature that it could only be formulated in counterfactual terms. A causal process was a process that *was capable of* transmitting a mark, so to say the light beam was a causal process was not to say that it was carrying any particular mark, but that, if it were marked, the mark would be transmitted.

Salmon decided in the 1990s to abandon the mark account, chiefly in response to criticisms from Kitcher (1989) and Dowe (1992). In its place he adopted a version of

---

[2] For further discussion of the relation between Salmon's mechanical philosophy and the New Mechanism see, e.g., Glennan (2002a); Williamson (2011).

Dowe's conserved quantity approach (Salmon 1994). On the Salmon-Dowe approach, a causal process is one that transmits a conserved quantity, while causal interactions consist of intersections of causal processes in which conserved quantities are exchanged. Salmon found the latter approach preferable in large part because it did not require counterfactuals.

The Salmon-Dowe theory is a theory of physical causation—a theory of what the biff is that courses through the basic physical constituents of the universe. As Dowe (2000) sees it, this empirical analysis is the second step in Canberra planning wherein we find the thing in this world that does the causal work. That analysis shows us that the biff is in the conserved quantities identified by physical theory.

For our purposes, the difficulty with the conserved quantity theory is not that it is wrong. Certainly, intersections of physical processes in which conserved quantities are exchanged are exemplary cases of causal interactions. The problem is that the theory reduces production to one thing (or more carefully a few things, corresponding to the set of conserved quantities recognized by physics). If the New Mechanist account is right, there are different kinds of production corresponding to different kinds of activities and interactions. Dowe and Salmon insist that when such "higher-level" producings occur, there will be physical causation "at the bottom," and we can grant that this is so. But without a better account of how these higher-level activities produce what they do, we shall not have an adequate theory of causation. This is what a New Mechanist account of production can provide.

7.2   Mechanistic Production and Relevance

Before turning to an account of what production and relevance are as relations in the world, I want to do a bit more conceptual analysis, exploring the role that concepts of production and relevance play in causal claims. In the previous chapter, we distinguished event causal claims from causal relevance claims. In the first case (C1) one event causes (or produces) another event, for instance:

Mary's examination caused Jane's fever.

Causal relevance claims (C6), on the other hand, do not assert something directly about productive relations between events, but indicate rather that some fact about or feature of the world made a difference to the occurrence of the event:

The weakened state of Jane's immune system was causally relevant to her death.

The weakened state of Jane's immune system is not an event; it is a standing condition upon which Jane's death depended. Had Jane's immune system not been weakened, then (we are imagining) the infection introduced by Mary's examination would not have ultimately produced her death.

While in ordinary causal discourse, some claims will describe productive relations, and others will describe relevance relations, it is possible to make causal claims that assert things about both production and relevance. They indicate both that some event causally produces another and say something about why this productive relationship holds. Consider the following examples:

(C8)   Mary's examination caused Jane's fever in virtue of the bacteria on Mary's hand.

Mary's examination caused Jane's fever in virtue of the lacerations in Jane's birth canal.

Mary's examination caused Jane's fever in virtue of a response from Jane's immune system.

Mary's examination caused Jane's fever in virtue of Mary's failure to wash her hands.

Mary's examination caused Jane's fever in virtue of Jane's prior skin contact with an infected patient.

All of these statements are of the form "c caused e in virtue of p" where c and e are events and p is any relevant feature of the causal situation. Making the kinds of relata explicit, we can see these sorts of statements as instances of a canonical form for singular causal statements.

Event c **produced** event e in virtue of **relevant feature p**.

I shall have more to say about the nature of events in a moment, but for now it is enough to understand events as concrete dated and located happenings, like a particular molecule binding to a particular receptor or two people meeting at the Café Central for coffee on June 1, 2016 at 9 a.m. While a description of these events may be abstract or incomplete, the productively related events themselves are fully concrete and determinate.

What though are features, and what are they features of? I have so far said they are features of "the causal situation." Briefly, I will take features to be any abstract characteristic or property of the entities, activities, and interaction or their organization that characterize the productively related events, intervening mechanisms, or their environment. Relevant features are those that make a difference to the occurrence of an event, while irrelevant features are those that do not. A simple example may help: if a ball strikes a window (an event) causing the window to break (another event), some features of the ball (e.g., its mass, velocity, and shape) will be relevant while others (e.g., its color or brand) will not. The list of statements about the causes of Jane's fever gives some idea of the variety of features that can be relevant to the occurrence of an event. It is notable that relevant features can include absences and omissions, like Mary's failure to wash her hands.[3]

---

[3] The account I have offered might naturally be taken as suggesting that causation should properly be understood as a three-place rather than two-place relationship. I have chosen to formulate MC as a two-place theory describing the productive relationship between events, but we might equally well think of the primary relation as a three-place one, with the two-place version being defined via quantification. If we

As our examples from the previous chapter suggest, many causal claims will make assertions about productivity without relevance or relevance without productivity. Sometimes we know that some circumstance is causally relevant to an event without knowing much about the particular events that were the immediate causes of the event. If Franz and Sisi get into a fight, we may not know what precipitated the fight, but we may know that Franz and Sisi have not been getting along, that Sisi has been depressed, or that Franz is a control freak, and we may have good reason to believe that any of these facts were relevant to the occurrence of the fight. Similarly, we may have good reason to believe that one event produced another without knowing what features of the event or surrounding circumstances made it so. If Franz walks into the room and Sisi rushes from the room, the proximity of the events gives us good reason to believe that one event produced (or contributed to the production of) the next—but we may equally have no idea about what features of Sisi's cognitive or emotional state (or Franz's appearance or behavior) were relevant to this occurrence.

The virtue of my proposed canonical form for causal statements is that it shows something about the relationship between productivity and relevance as causal concepts. On the analysis I am offering, productivity and relevance are not competing approaches to analyzing the nature of causal relationships, but rather are complementary concepts that refer to different features of the causal structure of the world.

In the rest of this chapter, I want to spell out just what these features are. To work out this account it will be helpful to focus the analysis around an example where the mechanisms involved are fairly well understood, but where the systems and processes involved are complicated enough to allow us to explore some of the complexities of causal relationships we seek to understand in science. So here, then, is a causal story:

Sisi walks into the St. Elmo's Steak House. The waiter Franz takes her order, a NY strip steak, medium rare. Franz gives the order to the kitchen, where Maria the line cook prepares the order. Maria puts a skillet on the stove to preheat. Meanwhile she takes a raw steak out of a container, pats it dry, rubs on a bit of olive oil, and seasons it with salt. When the skillet is hot, she puts the steak in the pan, searing the meat. After a couple of minutes she turns the meat, and then after a minute more puts the skillet in the oven to finish the cooking. When it is done, she plates the steak. Franz picks up the order and delivers it to Sisi. Sisi cuts the steak, takes a bite, and sighs with contentment.

This story describes a chain of productive causal relations between events (a single-case processual mechanism). According to MC, the existence of this processual mechanism

have a three-place causal relation $C(c,e,p)$, we can derive the two-place productive relation by quantifying over the third place in the relation $\exists x C(c,e,x)$. Similarly, we can derive the relation between causally relevant facts and effects by quantifying over the first place in the relation $\exists x C(x,e,p)$. Formally, such a proposal certainly works, as for any causing event there will be a multitude of facts relevant to the production of the effect, and for any causally relevant fact there will be a multiple set of causing events. Nonetheless, I have chosen to focus in MC on the productive relationship, as it is the standard one. However one construes the number of places in the relation, the crucial point is expressed in the definition in MC, namely that a productive relation between events depends upon the existence of a mechanism by which c contributes to the production of e.

entails that Sisi's ordering the steak, Maria's searing the steak, Franz's delivering the steak, and Sisi's taking a bite all contributed to the production of Sisi's contented sigh. Many features of these events and background conditions might be relevant to her sigh, from the amount of marbling in the steak to Franz's warm smile or the charming and historic bar at which Sisi is taking her dinner. Starting with this example, our task now is to understand in just what the production and relevance relations consist.

7.3 Events

To understand how one event produces another we must first have a clearer concept of what an event is. In keeping with our entity-activity ontology, we shall say that an event is just one or more entities engaging in an activity or interaction. For instance, Sisi's cutting the steak is an event. The activity is cutting, the entities required for the activity are Sisi, the knife and fork, and the steak. Because events involve entities engaging in activities and interactions, events are spatially localized particulars. Sisi's cutting the steak takes place at a certain place (the end of the bar at St. Elmo's) and time.

This definition of events excludes so-called "Cambridge events" (Kim 1974), which are events involving no direct activity of or change to an entity's intrinsic properties, but a change of state determined by relational properties. For example, if my wife gives birth to a son, that is an event in which she engages in activity, but my becoming a father is a Cambridge event, and is not, by itself, the sort of event that can produce other events.[4]

The entity-activity conception of event I am proposing bears resemblances to other more familiar accounts of events, but there are important differences. On the surface, the position appears close to Kim's property-exemplification view of events. Kim's view is that events are exemplifications of properties (or relations) by an object (or set of objects) at a time (Kim 1973; 1976). If exemplifying a property were the same as engaging in an activity, then the two views would coincide, but there are important differences between exemplifying properties and engaging in activities.

Properties are paradigmatically synchronic states of an entity that belong to that entity for some time. For instance, we might say of a cut of meat that it has a certain thickness or a certain percentage of fat. While such properties are difference-makers in cooking the steak, they are not activities and do not involve change. On our pre-theoretic conception of event, events are different from states, and do involve changes. To make his account work for events ordinarily conceived, Kim must take properties and relations to include ones that are "activity-like." For instance, Kim suggests that the event of Brutus stabbing Caesar involves the exemplification of the stabbing relation by the pair (Brutus, Caesar) at some time t.

---

[4] It is obvious that my becoming a father does in some sense cause lots of things, but what really causes these things is not the mere relational fact of my becoming a father, but the whole range of complicated psychological and social events that do involve actual happenings to me, which impact the life of me, my wife, my son, my insurance company, and so on—in so many ways.

Kim's move is consistent with a standard philosophical practice of equating events and states. But this maneuver reifies the activities and loses sight of the activity involved. The entity-activity conception, according to which the event of Brutus stabbing Caesar consists of three entities (Brutus, Caesar, the knife) engaging in an interaction (stabbing), better captures the temporally extended and dynamic character of the event.

Distinguishing events from states and properties is important for my account, because on that account only events (which involve activities) can be causally productive. States and properties like the thickness of the meat or the presence of oxygen in the kitchen count as events on Kim's view, but states like these cannot produce anything; they will often, however, be causally relevant. The steak's thickness, for instance, will be relevant to how cooked the center becomes during searing. In the case of the oxygen, while the presence of oxygen is a state of affairs in the kitchen that makes a difference to, say, the gas stove lighting, the presence per se does not produce anything. But it is only in virtue of this presence that there can be genuinely productive activities—namely the actual interactions of oxygen molecules with the gas in the process of combustion.

Another important difference between the entity-activity conception I am advocating and Kim's property exemplification view has to do with the "grain" of events. Kim takes a fine-grained view of events in which exemplifications of more or less determinate properties are distinct events. For instance, Kim claims that Brutus stabbing Caesar is a distinct event from Brutus killing Caesar, because killing is a different property than stabbing. An alternative, coarse-grained, view of events holds that the stabbing and the killing are the same event, and to refer to that event as a stabbing or a killing is simply to describe it at finer or coarser grains.

I shall take a coarser-grained view. In keeping with the account of entities and activities we have given, these things and the events they participate in are concrete particulars. Maria's cooking the steak is the very same event as Maria's searing the steak and finishing it in the oven, and it is also the very same event as the searing of that particular bit of Ferdinand the Bull's loin. Various descriptions may characterize the entities and activities more or less determinately, but they are composed of the same entities and activities, and so the same event.

On the other hand, even the maximally determinate activity does not necessarily encompass everything that is going on with the entities involved. Suppose, for instance, that as Maria sears the steak, she also sweats. The sweating is prima facie a different activity than the searing rather than a more determinate form of searing. Whether searing while sweating is a determinate form of searing—not searing simpliciter, but searing in a specific way—would depend upon whether the sweating is productively connected to the particular searing. If the sweat from Maria's brow drips upon the steak, and this sweat alters the chemistry of the searing, then this sweating is an aspect of activity that produces the cooked steak. If not, then it is a different activity and part of a different event.[5]

---

[5] The suggestion I am making here is not dissimilar to Davidson's (1969) causal criterion for individuating events. According to this criterion, an event is individuated by its causes and effects. If the sweating

While events are concrete particulars, different descriptions will pick out different features of those events, and those features may turn out to be causally relevant. More or less abstract descriptions of entities, activities, and events refer to more or less determinate features of those entities, activities, and events—but those features are real features, independent of our description of them. Consider, for instance, the steak: it is at once meat, beef, a bit of a Hereford Bull, and a bit of Ferdinand. These are not just different ways of describing the steak, but are ways of referring to different properties that the steak actually has. Similarly, we may say of Maria's cooking that it is cooking, searing, cooking medium rare, cooking to exactly 125 degrees Fahrenheit, etc.—and these descriptions pick out different but actual features of the concrete event. Any of these features may be causally relevant to an event's effects.

7.4 The Varieties of Mechanistic Production

According to the mechanistic account of causation, one event causes another when there exists a mechanism by which the causing event c contributes to the production of the effect event e. The task now is to spell out what "contributing to the production" consists in. In what follows, it will be useful to have some notation to remind us of the entity-activity characterization of an event. I shall continue to follow Craver's convention of using Greek letters to denote activities and capital Roman letters to denote entities, but I will use a functional notation, so that we denote the event of S engaging in activity ψ by $\psi(S)$. If an event is a monadic activity, like Sisi sighing, than the event is represented by a single place function, but if the event is an interaction between entities, like Franz picking up a plate, the function will have multiple arguments, e.g., $\Pi(F, P)$.

Productive relationships between events are compounded from productive relationships of the following types:

- **Constitutive production**: An event produces changes in the entities that are engaging in the activities and interactions that constitute the event.
- **Precipitating production**: An event contributes to the production of a different event by bringing about changes to its entities that precipitate a new event.
- **Chained production**: An event contributes to the production of another event via a chain of precipitatively productive events.

Using the metaphor of a causal chain, we can say that constitutive production is production of change within a single link of the chain, precipitating production is the production of change in one link by an adjacent link, and chained production is production of change by a first link on a remote link via a chain of intermediate links.

contributes to the searing, then there is one common effect and it is one activity. It differs from the yet coarser-grained approach he later adopted, which identified events with space-time regions. If it is possible for two different activities or interactions to be happening at the same place and time, then the space-time region criterion is too coarse.

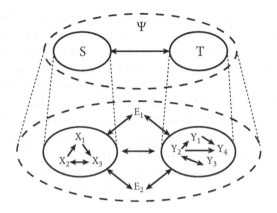

Key:

Roman letters represent entities; Greek letters represent activities and interactions; Solid lines are entity boundaries; Dashed lines are event boundaries; Dotted lines represent constitution relations.

E_1 and E_2 represent entities in environment. Interaction/activity labels are omitted from the lower half of the figure.

Figure 7.1   Constitutive production.

These three kinds of production are represented schematically in Figures 7.1–7.3.[6] I shall describe each type of production in turn.

Constitutive production: The most typical way of talking about event causation involves questions about how an event involving one object (e.g., throwing the ball) produces an event involving other objects (e.g., breaking the window), but our first step to understanding the nature of production is to consider the ways in which a single event produces changes in its constituent entities. As an example, let us consider our searing event $\psi(S, P)$. What makes this a productive activity is, in the first instance, the fact that the activity of searing produces changes in the steak and, to a lesser extent, the pan. Let us focus on the change that matters for the cook: searing both browns and cooks the steak; it also releases oil, water, and other compounds from the steak. How does the searing produce these changes?

Any proper account of how searing produces browning and other changes to the steak will necessarily refer to the constituents of which the steak is made. Those constituents are the muscle, connective, and fat tissues, and these tissues in turn consist

---

[6] These figures are generic representations of the varieties of production, so the elements and labeling will not correspond to the specific examples described in the text.

of cells, with these cells containing a variety of chemical constituents—principally water, proteins, carbohydrates, and fats. Cooking in general involves the breaking down and altering of these molecular constituents by the application of heat.

In searing (in a pan) the heat is transferred from the heated pan to the meat by conduction. This heat transfer initiates a variety of different chemical processes in the various parts of the meat. It melts fats; it denatures proteins, etc. The particular chemical properties that are characteristic of the activity of searing are those that cause meat to brown. These are broadly of two kinds. There are both caramelization reactions, which are simply reactions of sugars to heat, and Maillard reactions, which Harold McGee describes:[7]

The sequence begins with the reaction of a carbohydrate molecule (a free sugar or one bound up in a starch...) and an amino acid (free or part of a protein chain). An unstable intermediate structure is formed, and this then undergoes further changes, producing hundreds of different byproducts.... Maillard flavors are more complex and meaty than caramelized flavors, because the involvement of the amino acids adds nitrogen and sulfur atoms to the mix of carbon, hydrogen and oxygen, and produces new families of molecules and new aromatic dimensions.

(McGee 2004, loc 21618)

Notice that the changes to the steak produced by the searing depend upon changes to parts (and parts of parts) of the steak. The production of browning by contact with the hot skillet depends upon the interactions between carbohydrates and amino acids, interactions that occur via Maillard reactions, which occur only at temperatures above the boiling point of water. The searing also produces new entities. For instance, water and fat are released from the steak, with the water evaporating and the fat in its liquid form filling the pan. It is also important to observe the temporal and processual character of the event. The activities in which the parts of the steak and the pan are engaged all take time.

This form of production is constitutive because the changes in the properties of the entities arise from changes to the properties of its parts. The coloring of the steak is due to the refractive characteristics of the molecules produced by the Maillard reactions, so any account of how the searing produced the browning in the steak must refer to the way in which the heat-induced reactions produced new compounds *within* the parts of the steak.

While constitutive production is the production of changes to entities that are engaging in the activities that constitute the event, a description of the event will typically leave out some of the entities involved. Take as an example the event of Sisi ordering the steak. This event appears to be an interaction between Sisi and Franz, $\Omega(S,F)$, and indeed this ordering is a temporally extended process in which various parts of Sisi are enlisted in the production of certain utterances, and in which various parts of Franz are enlisted in the perception and interpretation of these utterances—where this

---

[7] My account of the chemistry of searing and other aspects of causal mechanisms involved in cooking is based upon Harold McGee's lucid discussions in his classic book *On Food and Cooking: The Science and Lore of the Kitchen* (2004).

perception and interpretation produces changes in Franz's cognitive state that are in turn productive of further links in the causal chain. However, this productive inter-action occurs via a mechanistic process that involves entities that are not part of either Sisi or Franz. Most obviously, when Sisi speaks to Franz, this interaction is mediated by an auditory signal that is transmitted via sound waves. Without the air, there would be no interaction between Sisi and Franz, and the activities of Sisi would not have pro-duced changes in Franz. Even in the case of searing, it is arguably the case that there are other entities involved besides the steak and the pan. For instance, the steak is brushed with oil and sprinkled with salt. The oil changes how heat is transferred from the pan to the meat, and the salt will alter the chemistry of the various heat-induced activities in searing. In general, fully concrete events involved in productive causal relations will typically involve more entities and activities than are referenced in any particular description of those activities. It is the fully concrete events, not any description of them, that are productive of changes in the entities involved in those events.

The examples I have discussed here are cases in which the constitutive production arises from an interaction between two (or more) entities, but monadic activities can be (constitutively) productive of changes to the entity itself. So for instance, the event of Sisi sighing is a monadic activity, in which activities and interactions between constituent parts of Sisi produce the sigh and also attendant changes in her facial expression, breathing, etc.

Precipitating production: While constitutive production is the way in which a single event, that is, a single set of entities engaging in some activity, produces changes in those entities, precipitating production is the way in which one or more events prod-uce another event. The idea of precipitating production is that one or more events trigger another event by creating start-up conditions for a different mechanism.

A simple example here involves the event of Maria (M) placing the steak (S) into the hot pan (P). Call this event $\Phi(M,S,P)$. This event itself is constitutively productive, in the sense that this event takes time and involves changes to the entities that are engaging in the activities, where these changes are produced by the activities and interactions of parts. The steak, for instance, can only be placed in the pan by Maria through the activities of her hands and arms.

But beyond itself being constitutively productive, the event $\Phi(M,S,P)$ also product-ively precipitates a distinct event, the searing event $\psi(S,P)$. It does so by producing the start-up conditions for the mechanism whose activity constitutes the event. The set of

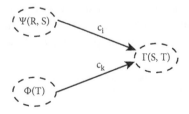

Figure 7.2 Precipitating production.

chemical activities and interactions that constitute searing require the surface temperature of the steak to rise above the boiling point of water, so until the steak is brought into proximity with a heat source such as the pan, it will not sear.

A generalized representation of precipitating production is provided in Figure 7.2. In this case, the precipitating conditions are represented as being produced by two previous events, which generate start-up conditions represented by c_i and c_k. These conditions could be of a variety of kinds. For instance, they could involve moving entities so that they come into contact with other entities, or specify changes in the state of entities in the precipitating events that will make them interact with other entities in other events, or creating new entities that are used in subsequent interactions.

Chained production: Events that precipitate other events via the relation of precipitating production form single links within causal chains. In Figure 7.2, we represented these links by arrows. Our story of Sisi ordering the steak describes a number of such precipitating events. Sisi's ordering (an interaction between her and Franz) brings about changes in Franz, which precipitate other activities on his part—walking to the kitchen, entering the order, etc.—which precipitate other activities, like Maria starting to prepare the order, and so forth. When an event c does not precipitate an event e directly, but instead precipitates other events, which precipitate other events in turn until one event in the chain precipitates the event e, we shall say that c and e are related by chained production. The general form of chained production is represented in Figure 7.3. Here we say that the event $\Phi_1(X_1)$ contributes to the production of event $\Phi_n(X_n)$ because there exists a chain of intervening events $\Phi_i(X_i)$, linked by precipitating production. Actual cases of chained production will seldom, if ever, involve single strands. That is why for one event to be a cause of another, it must contribute to the production of the second, rather than producing it outright. Events that contribute to the production of an effect will lie somewhere in one of the causal chains that ultimately precipitate the effect event.

In chained production the various activities and interactions in the causal chain can be of very different sorts. Sisi's ordering is one kind of production, while the steak's searing and Sisi's chewing, tasting, and smelling of the steak are all quite different kinds of production, because the activities (and entities) are quite different. Nonetheless, productive continuity of causal chains is maintained by these different forms of production. It is arguably only because of the heterogeneity of causal chains that we have need of the word "cause" at all. In a simple causal process, like the hammer nailing the nail into the wood, we can just speak of the activity of nailing. But in a complex causal chain, like the multi-stranded chain of events by which the Battle of Austerlitz led to Austria's signing a treaty with France, there are so many different sorts of entities and activities involved in getting from the battle to the treaty that the only economical

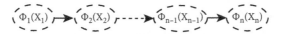

Figure 7.3 Chained production.

way to assert the existence of the productive relationship is with the generic language of cause and effect.

While precipitating production yields links in causal chains, the distinction between precipitating and chained production is not absolute. Since events themselves are entities engaging in activities, and these entities and activities are mechanism-dependent, events will consist of sequences or patterns of sub-events. Take, for instance, the event we have described as Sisi ordering the steak. In the simplest form, this interaction between Sisi and Franz can be decomposed into three events: Sisi speaks; that speech produces an audible signal that is transmitted through the air around Sisi; Franz hears the speech. These events in turn are further decomposable, so for instance Sisi's speech signal is comprised of a sequence of word vocalizations, which in turn are composed of phoneme sequences.

To conclude the discussion of the forms of mechanistic production, we can reflect briefly upon its relation to Salmon's account of production and transmission. My picture of the causal nexus, like Salmon's, sees it as consisting of a network of intersecting causal processes. The examples Salmon offered of production (the hammering of the nail or the pulling of the cart) are examples of the relationship I have called constitutive production, though the analysis is quite different from either the conserved quantity or the mark transmission theories. In place of Salmon's transmission, I have offered the account of precipitating and chained production. As with Salmon's analysis of transmission, the theory tells you how you get productive continuity between one event and another. The main difference is that the mechanistic account emphasizes how transmission of causal influence depends upon mechanistic processes, with their constituent entities and activities. Salmon might think my emphasis on how processes involve interactions of discrete parts is a mistake, as he admiringly quotes Venn's endorsement of the metaphor of the continuous rope of causation over the discrete links of the causal chain (Salmon 1981, 67). I myself have no problem with the metaphor of the rope, but it bears observing that ropes are only continuous entities if you do not look closely. Ropes are made by twisting fibers in bundles of greater thickness, and ultimately derive their strength and continuity from the interactions of their many parts.

With this basic account of the nature of mechanistic production in place, I will now try to address several objections that have been raised to productive theories generally and mechanistic theories in particular.

7.5   Fundamental Mechanisms and Production

In the mechanistic account of production, the productive capacities of wholes derive from the productive capacities of parts. The productive powers of causal processes derive from the productive continuity of the activities that constitute the stages within those processes. The causal powers of systems arise from the organized causal powers of their components. So, when Franz speaks to Sisi, the utterance is composed of words, the words of phonemes, and so on down to the physical motions of the air that are produced by Franz's vocal tract and which produce vibrations within Sisi's ear. And Franz's ability to say these things depends upon the various parts of Franz's vocal system—his mouth, his teeth, his diaphragm, his vocal chords, etc. This is just one example of a ubiquitous feature of the world's causal structure, the nesting of mechanisms within mechanisms that I call the mechanism-dependence of entities and activities.

It has sometimes been suggested that there is something circular in the mechanistic approach to causation (see, e.g., Psillos 2004; Craver 2007; Williamson 2011; Casini 2016). One event causes another just in case there exists a mechanism connecting them, but what it is to be a mechanism is to be a set of entities organized into causally productive activities and interactions. I believe there is no real ontological difficulty here: Higher-level and distal causal connections depend upon lower-level and proximate mechanisms; but this means that the activities and interactions that constitute the mechanism connecting c to e refer to different causal relationships than the one between c and e. In Glennan (1996) I offered an analogy between the composition of causes within mechanisms and the composition of sentences in predicate logic. In predicate logic, definitions of sentences are standardly given recursively, where we say, e.g., that, if p and q are sentences, then so are $p \, \& \, q$ and $p \to q$. Just as composition relations make sentences out of other sentences, mechanisms make causes out of other causes, but the recursive character of the characterizations in either case poses no problems for recognizing what counts as a sentence or what counts as a causally connected pair of events.

The analogy between composition of mechanisms and composition of sentences may break down in the base case. In predicate logic, there is a base case consisting of atomic sentences from which all other sentences are compounded. It is not, however, clear that there is an analogous base case for mechanisms. Perhaps there are basic causal interactions that can serve as a base case upon which productive causal relations may depend, but there may be other options. Perhaps, for instance, there is no end to the nesting, and there are mechanisms "all the way down."

Our discussion of this problem will be facilitated by some new terminology. Let us call a mechanism (that is, a set of entities engaging in activities and interactions) that does not depend upon other mechanisms, a *fundamental* (or *basic*) mechanism. All other mechanisms are *mechanism-dependent* or *compound* mechanisms. The following example should make the terminology clear. Assume that Newton's mechanics and theory of gravity were correct and that celestial bodies were point masses. If we consider the phenomenon of the motion of the earth relative to the sun, the solar system is a compound mechanism for the production of this phenomenon. It is compound because it consists of a number of the entities (minimally the planets and the sun) that engage in activities (attracting) with the other planets in the system. However, if gravity were a basic and non-mechanism-dependent force, the action of each single planet upon each would be produced by a fundamental mechanism.[8]

---

[8] In past work (Glennan 1996; 2002a; 2011) I have simply spoken of fundamental interactions rather than saying they are interactions produced by fundamental mechanisms, so this language represents a terminological shift. This shift does not affect the main substantive claim, which is that if there are such

The focus of this book has been upon compound mechanisms, but fundamental mechanisms still fall within the scope of the minimal mechanism definition, though as a degenerate case. In a fundamental mechanism, there are entities interacting, as when the planets exert forces on each other; the phenomenon produced is changes in the properties of the entities. The organization is limited to whatever relations between the basic entities influence the operation of the fundamental mechanism. So in our example the masses and relative positions of the planets determine how the fundamental gravitational mechanism operates. If there were fundamental mechanisms in the sense defined, then events in which a change to one entity produces a change to another via that fundamental mechanism would be related as cause and effect.

Perhaps the most obvious explanation of why we tend to think that there is a fundamental level is that the part-whole decompositions that we find in mechanistic science always come to an end. As Machamer, Darden, and Craver (2000) observed, mechanistic theorizing within any given scientific domain will assume some set of "bottom-out" entities and activities—basic building blocks whose reality and productive capacities are not in doubt and out of which mechanisms are built. For instance, within molecular genetics DNA is understood to be a double helix consisting of strands of nucleotides that contain the bases cytosine, guanine, adenine, and thymine (C, G, A, and T). The bottom-out entities are the nucleotides, and the bottom-out activities they can engage in include forming and breaking hydrogen bonds with complementary bases. Given these basic entities and activities, models of the mechanisms of protein synthesis can be constructed that show how these basic entities are organized into more complex structures that allow cells to synthesize various products.

Machamer, Darden, and Craver emphasized that what counts as a bottom-out entity or activity is domain specific. Practitioners in a domain will take as bottom-out entities and activities things that are known to be compounds. For instance, in the case under consideration it is understood that nucleotides consist of chains of atoms (and that these atoms and their bonding behavior are in turn determined by subatomic constituents).[9] So we can obviously not infer from the fact that there are bottom-out activities in a

---

fundamental mechanisms, there is a base case of mechanisms whose productive capacities are not themselves mechanism-dependent. This language is not without precedent. For instance, Salmon clearly thought about basic physical forces as "fundamental causal mechanisms" (Salmon 1984, 276). The advantage of saying these interactions are produced by fundamental mechanisms is that then every causally related pair of events will be connected by a mechanism.

[9] Although bottom-out entities and activities form the explanatory primitives within a given domain, the grounds for treating a set of entities and activities as bottom-out are epistemic and context-sensitive. The crucial question is whether it is possible to formulate an explanation of the phenomena one seeks to explain by appealing to those entities and activities. Empirical research often leads to the conclusion that a phenomenon cannot be accounted for without finding deeper structures. For instance, research in epigenetics has made it clear that gene expression depends in many contexts upon DNA methylation, which is the attachment of methyl groups to cytosine or adenine nucleotides. This fact about the mechanism entails that proper understanding of gene expression requires one not to take the base pairs C, G, A, and T as the bottom-out entities but instead to treat them as things with further parts, parts that engage in activities and interactions that explain the differing behaviors of methylated and unmethylated bases.

particular domain that there is a fundamental level of absolute bottom-out entities and activities. At best, we can argue that physical theory will, at any given time, have identified some set of entities and activities as the most fundamental known, and that such a set of entities and activities might turn out to constitute a genuine ontologically fundamental level.

The idea that there is a fundamental level has its philosophical roots in ancient atomism. The view was central to the mechanical philosophy of the seventeenth century. Some mechanists, like Newton and Boyle, held that the world was composed of a set of fundamental particles—microscopic corpuscles—and that all physical phenomena were ultimately the product of basic mechanical interactions between these corpuscles. Since the seventeenth century, there have evidently been changes in our understanding of what the atoms are and what properties they have, but contemporary versions of this doctrine, now generally called microphysicalism, owe much to the seventeenth-century account.[10]

Philip Pettit defines microphysicalism as "the doctrine that actually (but not necessarily) everything non-microphysical is composed out of microphysical entities and is governed by microphysical laws" (Pettit 1993, 253). For Pettit and many others who defend microphysicalism, the reason to accept microphysicalism is that without it we cannot defend physicalism. While the meaning of physicalism is itself a matter of much debate, the basic idea of physicalism is that everything in the world is physical, in the sense of being made of the stuff physics describes, and thus everything is ultimately governed by the laws of that stuff. Pettit and others worry that absent a microphysical level to which to tie things, the claim that everything is ultimately physical will lose its meaning.[11]

In some of my earlier discussions of causation (Glennan 1996; 2011), I have been moved by similar considerations, which can be put in the form of an argument:

There exist productive causal relationships in the world.

If there is no level of fundamental entities whose interactions are productive causal relationships, then there will be no productive causal relationships at all.

Therefore there must be a level of fundamental entities.

---

[10] For a brief historical account of seventeenth-century Mechanism and its relation to the New Mechanism see Craver and Darden (2005). For some historical accounts of early Mechanism see, e.g., Boas (1952); Dijksterhuis (1961); DesChene (2001); Roux (2018).

[11] Not all philosophers would agree microphysicalism is required for physicalism. Hüttemann and Papineau have argued that it is important to distinguish what they call levels physicalism from part-whole physicalism (Hüttemann and Papineau 2005; Papineau 2008; Hüttemann 2010). Both find levels physicalism plausible but microphysicalism false. While there are certainly problems with the microphysicalist picture of the fundamental level—see Schaffer (2003); Ladyman and Ross (2007); and my comments below—the mechanistic account of production depends upon a notion of mechanism-dependence, which is a species of part-whole dependence, so if we are to give an account of mechanistic production we need to get some understanding of how these relations bottom out.

In retrospect, I think that this conclusion is too hasty. As an empirical matter, I do not doubt the first premise, but the second premise is hardly self-evident. If there were such a fundamental level, it could ground productive causal relationships, but there may be other ways to ground them. It seems at least conceptually possible that there is a never-ending series of decompositions of wholes into parts—all with mechanism-dependent interactions connecting them. We cannot rule out a priori the possibility of mechanisms all the way down.[12]

But suppose for the moment that the microphysicalist assumption that there is a fundamental level consisting of atomic entities is correct. There remain further questions about the kinds of relationships that would obtain between entities at this level, and how these relationships should affect our understanding of causal production within mechanisms. As we saw in Pettit's formulation, the most common supposition is that there will be some set of fundamental particles, and the interactions between these particles (and thereby the behavior of everything composed of them) will be governed by some set of physical laws. But just what it is to be governed by a law is a notoriously difficult question. Here are three possible interpretations:[13]

- Humean Laws—Interactions are nothing more than instances of patterns and law statements are simply descriptions of these patterns.
- Real Laws—Laws are genuine things in the world that govern interactions and explain regularities.
- Causal Powers—Interactions occur in virtue of the causal powers of individual entities involved in the interactions, and law statements are generalizations that describe patterns in the exercise of these powers.

We can think of these three interpretations as three different accounts of the truthmakers for claims about interactions between fundamental particles.

On the Humean view, there really are no laws (qua things in the world). Laws are nothing more than the most powerful and economical descriptions of the patterns found in the world. If the Humean view were right, it would undermine the account of causal productivity that I have proposed. The grounding of the most fundamental causal claims turns out not to be intrinsic to the cause and effect, but is relationally

---

[12] There is also what David Lewis has called the atomless gunk hypothesis. I do not think this can be ruled out a priori, but if there is atomless gunk, I think the whole of the gunk in some sense is a fundamental level. The fact that the gunk is divisible (in the sense that we can separate gunk into regions) is no more a sign that there are parts of the gunk than the fact that we could split the pat of butter into any number of pieces is a sign that the pat of butter has parts.

[13] These possibilities are discussed in more detail in Glennan (2011). The Humean Laws view is associated primarily with the Mill-Ramsey-Lewis account of laws, while the Real Laws position is most prominently associated with the Armstrong-Dretske-Tooley view of laws. What I am calling the powers view is close to the view of powers espoused by Cartwright (1989; 1999), and more recently by Heil (2012). My account differs principally in its insistence of understanding the powers of compound systems (Glennan 1997b). For another useful way of taxonomizing the views about the relation between laws, powers, and causes see Tooley (2009).

grounded in the pattern. From a causal realist's perspective, the Humean view is a view that simply denies both that there are causes and that laws do or govern anything.

On the other hand, both the real laws and the causal powers views can provide grounding for the productivity of higher-level causes. Both views offer an account of an oomph-giving relation between the bottom-out objects, but they differ on the source of the oomph. John Heil characterizes the difference between the powers and real laws view as a difference between what he calls internalist and externalist conceptions of law:

An externalist conception of laws makes laws out to be entities in the universe... in addition to propertied objects... An internalist conception, a conception that regards properties as powers, encourages the thought that laws are more aptly regarded as linguistic items: equations, formulae, or generalizations that are meant in effect to codify the contribution made by particular properties to the dispositional makeup of their possessors.... To a first approximation, externalists think of laws as governing objects and holding under "ideal" circumstances; internalists think of objects as self-governing and law statements as attempts to distill the contribution particular kinds of property make to objects' capacities. (Heil 2012, 99)

If there is a fundamental level of entities and interactions, the real laws view would seem to have the advantage of explaining patterns of behavior among these entities. The idea is that laws express relations between properties (construed as universals) and the fact that entities behave in accordance with these laws (like the behavior of charged particles in magnetic fields) is explained by the fact that their behavior is governed by them. On the causal powers view, there is no explanation of the uniform behavior of interacting entities, because there is no governing law that is external to it. Still the causal powers view has the advantage that it provides only what is minimally necessary to get causal productivity without positing some new metaphysical entity. It does not seem to be a conceptual truth that determination requires instantiation of a universal. It seems possible that one thing can determine another without instantiating a law. This is Anscombe's intuition.

There is, moreover, a natural alliance between the powers view of fundamental interactions and the mechanistic approach. I have argued (Section 2.6) that non-fundamental laws are appropriately understood as generalizations describing the behavior of mechanisms, and that the real source of causal power within mechanisms is in the particular entities, activities, and interactions that constitute the mechanism. The power is within the mechanism and not external to it; it is self-governing. The argument for the self-governing character of compound mechanisms is not available for fundamental mechanisms, since I cannot say that laws describing the behavior of fundamental mechanisms depend upon entities and activities in lower-level mechanisms. Nonetheless, if fundamental entities have their powers internally rather than externally then we would get internalism and singularism from top to bottom. Conversely, if the real laws view obtains at the fundamental level, there is a sense in which nothing is self-governing; if God moves the fundamental entities, God moves everything. So, while I cannot rule out the real laws view any more than I can the

Humean laws view, the powers view seems to me the view that provides the simplest and most coherent account of laws and interactions.

This whole argument presupposes microphysicalism and the idea that there is a fundamental level, but I have already suggested that this assumption is open to question. For one thing, there may not exist a fundamental level at all. It may be the case that there are no metaphysical atoms, and instead that there are mechanisms (or something else perhaps) all the way down. More importantly, even if there is a fundamental level, it may turn out that microphysicalism's assumptions about what the denizens of that level are like may be misguided.

The microphysicalist picture has strong affinities with the seventeenth-century mechanical philosophy, and much has happened to our understanding of the nature of physical reality since that time. In some cases, a contemporary microphysicalist can easily adjust their account to conform to these advances. For instance, it is easy to update the list of fundamental properties to drop size and shape and to add charge and spin. It is also possible to relax assumptions about regularities at the fundamental level to allow for indeterministic interactions. However, even with these amendments, microphysicalism supposes that the physics of the fundamental level is a classical physics. I will not hazard a precise definition of what would make a physical theory classical, but at the least it is fair to say that microphysicalism shares with classical physics the idea that if there is a fundamental level, there will be some set of identifiable objects at that level that have definite locations, boundaries, and other properties— properties whose values are determinate in advance of measurement.

Quantum mechanics suggests that the world is not this way. The entities described by quantum mechanics do not behave in object-like ways. They do not always have definite locations, and may be spread across the universe until their position is measured. There are also classically inexplicable limits to measurement (the Heisenberg uncertainty relation). Finally, and most importantly, there is the phenomenon of entanglement. Quantum systems obey non-classical laws of composition which make it impossible to characterize the state of composite systems wholly in terms of the states of their parts. Quantum entanglement may undermine the very conception of part that seems essential to the mechanistic viewpoint. Moreover, the problem cannot be contained within the quantum domain, because systems of any size are composites of quantum systems. It is also not reasonable to think that these phenomena are artifacts of an incomplete or possibly incorrect theory. Non-classical entanglement is an empirical fact that can be produced in the laboratory, and Bell-theorem results argue against interpreting non-classical features of quantum systems as reflecting incomplete theories.

A scientifically informed metaphysics should not make ontological suppositions that ignore these non-classical features of quantum mechanics. We should not assume, just because it conforms with our intuitions, that nature is ultimately grounded by a fundamental level of "little things and micro-bangings."[14] At the same time, the

[14] This phrase is used by Ladyman and Ross (2007) to derisively describe what they call "scholastic metaphysics," of which microphysicalism is an exemplary case. While I shall not discuss their positive

naturalistic approach cannot ignore the explanatory successes and apparent ontological insights arising from areas of science outside the quantum realm. The idea that phenomena are produced by the organized activities and interactions of entities that are parts of larger wholes is not just an idea that we draw from our folk conception of the world or from metaphysical doctrines like atomism; this is a persistent feature of scientific theories across the special sciences, and also in parts of physics—that snails and soils and solar systems and cells have parts is settled science. Our task, then, is to reconcile these two facts about the world that we take from our understanding of the fruits of science.

One of the things we must admit about the non-classical features of quantum mechanics is that they seldom manifest themselves above the atomic level of organization. Quantum indeterminacy and entanglement do not appear, for instance, to play a significant role in the explanation of the behavior of cells.[15] Probabilities, including objective probabilities, can characterize the behavior of entities, activities, and mechanisms, but these do not require quantum indeterminacy (Section 5.4). What this suggests is that, in some way, out of the non-classical world described by quantum mechanics, classical phenomena emerge; how this happens is something philosophers and physicists should seek to account for.

Meinard Kuhlmann and I have argued that such an account may be found in the theory of quantum decoherence.[16] Decoherence theory seeks to explain how, under certain conditions, quantum mechanics predicts the emergence of approximately classical objects with approximately classical properties. The term "coherence" (or entanglement) refers to the degree of (coherent) superposition between states of a system, and decoherence refers to a process by which the coherence of these states is lost. Coherence may be thought of as a measure of "quantumness," so the more a system decoheres, the more classical it becomes. A familiar example of a coherent state is the state of an electron traveling through a double-slit apparatus. Travelling through the double slit, the classical supposition is that the particle must go either through one slit or the other, but quantum theory predicts that there will be a coherent superposition of passage through the top slit and passage through the bottom slit, and experiment confirms a wave-like interference pattern created by this superposition.

Decoherence theory helps explain why coherent superpositions like those created by the double slit are so rarely detectable beyond atomic scales. The key insight behind the theory is that macroscopic systems are not closed, and are constantly interacting

---

ontological account (ontic structural realism), my attempt in this chapter to explain how mechanistic ontology can dovetail with quantum mechanical ontology is in keeping with their principle that any metaphysical account must be constrained by our best current theories of physics.

[15] Some scientists and philosophers have argued that non-classical features of quantum mechanics may be relevant to the explanation of some biological phenomena (e.g., protein folding), but this work is highly speculative. For discussion see Davies (2004).

[16] Kuhlmann and Glennan (2014) provide a reasonably non-technical introduction to decoherence theory along with a discussion of its implications for our understanding of the relationship between quantum mechanical and mechanistic ontologies. For a general discussion of decoherence theory and the emergence of classical phenomena see Zurek (2003).

with their environment. The theory shows how these interactions will cause a coherent (non-classical) system to decohere. The constant interactions with the environment do not eliminate the entanglement, but rather spread it out throughout the environment so that the entanglement is no longer locally perceptible. While decoherence theory does not solve the measurement problem, it does go some way toward explaining why the world that we observe behaves classically.

If there are not mechanisms all the way down, a mechanistic account of causal production will need some kind of bottom-out level of classical objects that can engage in productive interactions, and decoherence theory explains the conditions that will lead to the appearance of such a level. In that sense, decoherence gives an account of the transition from quantum systems to a "fundamental classical level."

But the theory also shows why talk of *the* fundamental classical level is misleading. The reason is that there is no particular sort of entity or particular scale at which the world becomes classical. While it is typically the case that larger objects will be more entangled with their environment and hence more classical, there is no one story to be told. It is possible in special circumstances to maintain coherent states in very large systems (e.g., the supercollider at Cern). On the decoherence account, the emergence of classicality is a local affair; it occurs whenever the systems interact to a significant degree with their environment.

The piecemeal character of the emergence of classicality that is suggested by decoherence theory is at odds with most philosophical accounts of microphysicalism, since these typically suggest that there is *a* fundamental level of fundamental particles governed by fundamental laws; but the decoherence account works well for the mechanist. Even if there is no general story about the emergence of classicality, within any particular system or class of systems, there will be some point at which quantum effects effectively disappear, permitting us to detect and manipulate objects that behave and interact with other objects in an approximately classical way. These local entities and interactions will form the bottom-out or fundamental classical level that grounds mechanistic production.

The fundamental classical level is not absolutely fundamental, since the entities at this level will be composite systems, and it will be able to explain capacities of the whole system in terms of their parts. The difficulty is that the parts we find are not classical objects. At this level we reach the end of classical mechanical explanation, but not the end of explanation overall.[17]

---

[17] I have described the situation, the quantum/classical boundary represents the end of mechanistic explanation, but the actual situation is less clear. If one relaxes certain assumptions about what kinds of entities may count as parts of mechanisms, certain quantum mechanical explanations will be amenable to (non-classical) mechanistic explanation. As we argue, once one recognizes that the primary mode of organization within a mechanism is causal rather than spatial, and that the entities involved do not need to be physical objects with clear spatial locations, then we see it is possible to explain at least some quantum mechanical phenomena mechanistically. See Kuhlmann and Glennan (2014), pt. 5; Kuhlmann (2015).

Let me try to summarize the conclusions we can draw from this section. We began with the observation that the mechanistic approach to production involves a part-whole strategy in which the productive power of larger entities and processes derives from the parts of those entities and processes. Given this account, it might seem natural to believe that there would be no productive powers (no biff) at all, unless there were set of fundamental mechanisms—a set of productive relationships that obtain between a set of basic particles or objects. The claim that there is such a level is roughly equivalent to the doctrine commonly known as microphysicalism.

I have argued that notwithstanding any intuitions we have about the necessity of grounding, there are not decisive philosophical arguments against the possibility of mechanisms all the way down. But the more important concern for the New Mechanism is not some hypothetical worry about not being able to find a fundamental level, but rather the known ontological features of quantum systems. Mechanistic explanations as well as the fundamental level assumed by microphysicalism take a classical view of objects and their interactions, and in quantum mechanics some of the very ideas of classical objects and classical modes of composition break down. So our real concern must be how classical objects, properties, and interactions can emerge out of microsystems consisting of non-classical parts. However it works, the success of mechanistic science leaves little doubt that this emergence does occur, and the theory of quantum decoherence provides a promising start on gaining a theoretical understanding of why this is so. We are left with the idea that there are structures in reality at micro-scales that are not amenable (or at least fully amenable) to mechanistic analysis, but that they can provide grounding for a fundamental classical level from which larger compound mechanisms may derive their productive powers.

7.6 The Problem of Irrelevant Production

The New Mechanist account of causation falls within a family of geometrical/mechanical approaches that emphasize the intrinsic character of causal relations and the requirement of productive continuity of causal processes. Advocates of difference-making approaches to causation argue that all such processes fall prey to two kinds of problems. The first, irrelevant production, involves cases where a pair of events that are intuitively causally irrelevant to each other are nonetheless connected by a causal process. The second problem, non-productive causation, involves pairs of events that intuitively are causally related, but where there is no connecting process. In this and the following section I will argue that these problems are indeed problems for older causal process theories, but that the account of production and relevance developed here has distinctive features that allow it to avoid them.

The problem of irrelevant production is simply this: given a characterization of causal production in terms of some notion of basic physical influence or biff, as for instance one finds in the conserved quantity theory, it will be the case that many events are productively related to events that they are nonetheless causally irrelevant to. The original version of this objection is from Hitchcock (1995), but I quote here a concise version from Woodward:

> Suppose blue chalk is applied to the tip of a cue stick, which is used to strike the cue ball which in turn strikes the eight ball, which goes into the pocket because of the impact of the cue ball. Portions of the chalk are transmitted to the cue ball, and from there to the eight ball. There is a mechanical relationship in the paradigmatic sense of spatio-temporally continuous transmission of energy/momentum between the presence of the chalk on the cue (P) and the falling of the eight ball (F), yet P does not cause F in the difference-making sense.   (Woodward 2011, 414)

Notice how Woodward characterizes the productive or causal/mechanical relationship in terms of the transfer of energy/momentum. If this is what production comes down to, then Woodward would be right that one can have production without difference-making; but on the mechanistic account of production, these are not productive interactions, at least not productive of the effect whose causes we are seeking, which is the eight ball falling in the corner pocket.

On the New Mechanist account, there is no one characteristic of productive interactions—but rather specific kinds of production coming from specific kinds of activities. The chalking of the cue is an activity that produces a change in the cue, covering it with particles of chalk. When the chalked cue strikes the cue ball, there are two distinct activities going on—what we might call striking and coloring. Striking is the activity whereby energy and momentum are transferred from the cue to the ball, while coloring is the activity whereby chalk particles are transferred from the cue to the ball. The different activities produce different kinds of changes, which will be relevant to different kinds of phenomena. It is very much the same as in our example of ordering the steak. Searing produces heating and coloring, evaporates water, melts fat, etc. Ordering the steak produces entirely different sorts of changes. It does not heat or brown the waiter, but rather changes his mental state.

The New Mechanist account of production is an account of high-level production, in the sense that it grants that the sorts of activities and interactions described in sciences like chemistry, biology, and economics can produce changes. It also suggests that non-scientific concepts like scraping or cutting or ordering generally characterize activities that produce changes in the entities involved in them.

At the same time, the account does not deny that the productive character of these higher-level activities depends upon productive relations at lower levels of mechanistic organization. Our analysis of constitutive production shows in fact that the way in which an interaction produces changes in the interactors is by productive activities of their components. For instance, the way in which searing produces browning is via, among other things, Maillard reactions, which are productive interactions among molecular components of the food that is being browned. This means that ultimately the productive character of higher-level activities may depend upon physical interactions of the sort that Dowe has attempted to characterize. For instance, the chemical transformations in the Maillard reactions will certainly obey conservation laws, and we will be able to observe consequences of this at higher levels. Searing will reduce the mass of the steak, mostly because of the evaporation and release of retained water and the melting and release of fats. These materials are transferred into the atmosphere and into the pan, so total mass is conserved.

Even though the mechanistic account grants the dependence of higher-level production on productive relations between parts, the lower-level relations are only productive insofar as they are part of the mechanisms responsible for the higher-level production. So, for instance, both the activities of striking and coloring of the cue ball will depend upon lower-level mechanisms, but it is only the lower-level mechanisms involved in striking that account for the production of the change in the ball's velocity and momentum.

Here, though, is an objection: Events themselves (as opposed to their descriptions) are concrete, fully determinate with respect to every detail, down to the most fundamental physical entities and interactions. Surely the chalk on the ball produces some changes to the concrete event of the cue ball going into the pocket, even if it does not make a difference to its going into the pocket. If this is so, is it not the case that the chalking is productively related to the going into the pocket?

In response, we must make two points. First, we may have been too hasty in assuming that the chalking does not make a difference to the eight ball going into the corner pocket. Pool players chalk cues for a good reason. The chalk on the cue's tip increases the grip of the cue tip on the ball and allows the player to impart spin to the ball. This spin can change how the cue ball strikes the eight ball, and thereby influence whether the cue ball goes in the pocket. Of course, it is not the color of the chalk that alters the interaction with the cue ball but rather the texture of the chalk, so the coloring does not productively cause the ball to go in the pocket, but the chalking might. It even might cause it by some other mechanisms. For instance, chalking may be part of the pool player's routine, and the accuracy of the striking may depend psychologically and physiologically upon this routine. If this is so, then the chalking may be productively related to the ball going into the pocket.

The second, and more important, point is this: While lots of "irrelevant" interactions may contribute to the production of the fully concrete event, these productive interactions fail to make a difference to an event abstractly described. The concept of difference-making is connected in the first instance not to completely concrete events, but rather to higher-level events, or, alternatively to abstract features of the concrete events. When one identifies or describes an event, one invariably does so at some level of abstraction. In characterizing the concrete event of the ball falling in the pocket as "the ball falling in the pocket" one has abstracted from concrete features of the event— like the exact point and angle of entry and speed at which the ball falls into the pocket. When one says that the chalking does not make a difference to the ball falling into the pocket, one is not saying it makes no difference to the event at all, but only that the chalking did not make a difference to the ball falling into the pocket, which is an abstract feature of the ball's interaction with the pocket.

7.7 The Problem of Non-Productive Causation

While the problem of irrelevant production questions the sufficiency of production for causation, the problem of non-productive causation questions its necessity. Many philosophers believe that causes can make a difference without being linked via productive interactions to their effects. The simplest such cases involve omission and prevention. For instance, we might say that Maria's failure to remove the pan from the burner caused the steak to burn, or we might say that Max the chef's timely reminder to Maria to take the pan off the burner prevented the steak from burning.

Upon reflection, it is not, however, obvious that these are cases of causal dependence at all. Clearly in the first instance the burning counterfactually depends upon Maria's failure and in the second case the absence of burning counterfactually depends upon Max's reminder, but can an omission (the failure to take the pan off the burner) *cause* anything, or can a prevention (the absence of burning) be an *effect*?

Phil Dowe (2000; 2001) believes that omission and prevention clearly are not cases of genuine causation. He treats them instead as instances of what he calls "quasi-causation." The basic idea of quasi-causation is that quasi-causation is "possible causation." There are various different sorts of quasi-causal relations, but let us consider just one, simple omission. Dowe proposes:

Not-A quasi-caused B if B occurred and A did not, and there occurred an x such that (O1) x caused B [and] (O2) if A had occurred then A would have prevented B by interacting with x.
(Dowe 2001, 222)

In the case of Maria's omission, B is the burned steak, x is the interaction between the steak, pan, and burner that caused the burning, and A is the removing of the pan from the burner that Maria actually failed to carry out.

Dowe argues that, because quasi-causation refers to possible genuine causation, we can, for many practical purposes, treat them as the same. For instance, quasi-causes are explanatorily relevant and quasi-causal responsibility can be sufficient for moral responsibility. What Dowe calls "the intuition of difference" (218) explains cases like Maria's burning the steak, in which, upon reflection, we see that her failure to take the steak off the burner did not literally cause the burning, even though it explains the burning, and even though the manager can hold Maria responsible.

On the other hand, Dowe acknowledges that the intuition of difference does not hold sway in all cases, and that sometimes we have what he calls the "genuinist intuition." Here he cites Michael McDermott's example: "chopping off someone's head causes death (if anything causes anything), but does so as a prevention, since chopping off the head prevents processes which would have caused the person to continue living" (218). Chopping off the head is an instance of what Schaffer has dubbed "causation by disconnection" (Schaffer 2000), and it is only a particularly graphic example of a phenomenon that appears to be quite widespread.[18]

One option here is to bite the bullet and argue that events like chopping off heads do not really cause things. This is what Dowe recommends (though he argues that his quasi-causation account provides some comfort for those afflicted with the genuinist intuition). Alternatively, one may bite a different bullet and give up on production accounts of causation altogether, as Woodward would advise. I would like to pursue a middle course that attempts to really honor both intuitions. That course suggests that cases of omission and prevention are not cases of genuine causal production, but cases of disconnection are. Since cases of disconnections are typically analyzed as cases of double prevention, this will take some explaining.

First, though, let us consider why omissions and preventions cannot productively cause anything. The reason is straightforward. Omissions and preventions are not events. They are not entities engaging in activities, and hence they are not producers of change. In the case of Maria's burning the steak, the burning of the steak is produced by the continued heating of the steak, and this was produced in part by Maria's putting the steak on, but Maria's failure to take it off is not an activity, but a non-activity. Similarly, we may argue that Max's reminder produced Maria's removing of the pan from the burner, but it did not produce the non-burning of the steak—because the absence of burning is not an event.

But if prevention relations are non-causal, then how can disconnection (i.e., double prevention) cause anything? To explore this problem, let us change our restaurant story a bit. Suppose that Sisi has a hankering for wild boar. Franz, who is quite fond of Sisi, knows that there is no boar meat in the restaurant, but quickly heads out to find and shoot one in the nearby forest. Raising his rifle, he pulls the trigger, shoots and kills the boar, and (after a complicated causal chain) delivers the boar steak to the happy Sisi. My claim is that Franz's shooting the boar causes (in the sense of produces) the boar's death, and then by a sequence of other productive causal relations, the arrival of the boar steak at Sisi's table.

The problem with this story is that there are many disconnections involved. Let us focus on a well-known one discussed by Schaffer—the firing of the gun. Let us suppose that Franz's gun uses the bolt-action firing mechanism in Figure 7.4.

The mechanism works in this way: The bolt action that loads the round also cocks the firing pin, which is to say that it pulls the firing pin backwards against the resistance of a spring. The firing pin is held in place by a flange on the dark piece in the middle of the diagram (labeled A), which is called the sear. When Franz pulls the trigger, the trigger pushes up on the sear, dropping the flange and releasing the bolt. The potential energy stored in the coiled spring is released, accelerating the firing pin toward the

---

[18] The fact that so many cellular and neurological processes involve blocking and inhibition was in fact what moved Craver (2007) to adopt Woodward's manipulability approach to causation.

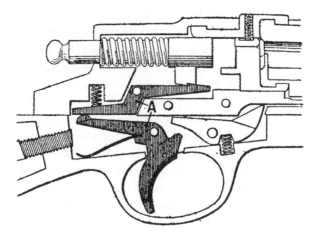

Figure 7.4   Disconnection in a trigger mechanism.

primer, igniting the primer, which in turn ignites the powder with an explosive force that pushes the bullet out of the barrel.

This is an example of disconnecting: pulling the spring does not transfer energy or momentum to the bolt, it simply disconnects it from the sear, which was preventing its release. According to this account, the only production here is caused by the spring. Because of this, Schaffer's contention is that on production or transference accounts (what he calls the "physical or intrinsic connection approaches"), pulling the trigger does not produce and therefore does not cause the firing of the gun.

On the basis of this and other examples, Schaffer concludes that "causation by disconnection is causation full force . . . Causation by disconnection is different than causation by connection (and liable to be overlooked if one concentrates on colliding billiard balls) but no less causal thereby. There is more than one way to wire a causal mechanism" (2000, 289). I agree with Schaffer that we cannot count cases such as Franz pulling the trigger as anything less than full-fledged causes. The way to do this, however, is to argue that on the New Mechanist approach to production, pulling the trigger actually produced the firing.

Here is how we do it: The gun has a certain capacity—the capacity to fire a bullet— that is straightforwardly mechanism-dependent. When Franz pulls the trigger, he produces a change in the state of the gun that precipitates (or "triggers") this capacity. When that capacity is triggered, the gun—and the whole gun—fires the bullet. It does so by the activities of its parts, in particular by the acceleration of the firing pin toward the primer, etc. In claiming that there is a productive relationship between pulling the trigger and firing the gun, I am not calling into doubt Schaffer's claim that this is causation by disconnection. What I am instead claiming is that the disconnection is one stage in the mechanistic process by which pulling the trigger produces the firing. We might call this "production by disconnection."

The reason that pulling the trigger can be said to produce the firing on the New Mechanist approach where it cannot on the physical connection approach has to do with differing understandings of what we might call the "width" of the productive process. The sear does not transfer energy to the firing pin, so if we focus on these pieces of the mechanism alone, we see only disconnection. The production, though, is not the work of the trigger and the sear by itself, but of the whole mechanism that is the gun. The productive capacity of the gun as a whole turns out to depend not only upon connections, but also upon disconnections of its parts. That is, as Schaffer says, how the mechanism is wired.

A defender of the physical connection account of production could perhaps insist that pulling the trigger only quasi-caused the firing, and that our assertion must be understood as contrasting the actual process in which the firing occurred with a non-actual process in which it did not. But I do not think that this relationship must be understood in terms of counterfactual dependence. There is an actual mechanism, and with it an actual causal process by which trigger-pulling fires guns. The relation between these events is intrinsic and non-contrastive, though the intrinsic connection runs not just from trigger to bullet, but through the mechanism of the gun as a whole.

Let us compare this case with a case of double-prevention from Ned Hall (2004, 241–2). Double preventions look a lot like disconnections and Schaffer treats them as the same, though I will argue they have a somewhat different character. It is World War III and Suzy is flying her bomber toward the target. Billy is flying a fighter. Billy shoots down Enemy, who otherwise would have shot down Suzy. Suzy drops her bomb on target. Billy's role is to prevent the preventer Enemy from interfering with Suzy—hence double prevention. This case appears structurally similar to the case with the gun. Suzy's bomber is like the firing pin, her dropping the bomb like striking the primer charge, Enemy is like the sear holding Suzy up, and Billy is like the trigger, who by removing Enemy, enables Suzy to drop the bomb. In Hall's interpretation of this situation, Billy causes the bombing in one sense; It depends (counterfactually) on his shooting down Enemy. In the other sense of cause, though, he does not, because there is no connecting process between Billy's shooting down Enemy and Suzy dropping the bomb.

How shall we interpret this case from the New Mechanistic point of view? One way we might do this is as follows: Thinks of the air force (or more specifically the planes on this mission) as a mechanical system like the gun. The air force drops bombs in much the same way that the gun fires bullets. The various planes flying on the mission are parts of this system, and it is the system as a whole, including the escorts, the signal jammers, etc., that produces the bombing. This picture seems right so far. Air forces really are systems that produce phenomena by means of the organized activities and interactions of their parts. That is to say, they are mechanisms, and what they produce, among other things, are bombings.

But what of the analogy between Billy and the trigger? Does Billy shooting down Enemy produce the bombing in the same way that Franz pulling the trigger produced the firing? I think not, for two reasons. Billy shooting down Enemy plays a different kind of role than Franz pulling the trigger. While Franz pulling the trigger initiates or precipitates the firing, Billy shooting down Enemy does not initiate or precipitate the bombing (which is already going on), but merely prevents it from being interrupted. This preventing it from being interrupted (or double prevention) is a kind of shielding of the mechanism against interference, rather than a triggering of the mechanism. If one were to look for something truly analogous to pulling the trigger on the gun, we would have to look elsewhere—perhaps to General, who launches the air force strike by giving the order. Conversely, if one were looking for a real analog to Billy in the gun case, one would have to add a preventer that needs to be shielded. Suppose, for instance, that the firing mechanism on Franz's gun was very sensitive to damp. If water splashes on the top of the barrel it leaks in and ruins the primer. To make sure the bullet fires, then, Franz always goes boar hunting with his umbrella, which prevents the water from preventing the primer from exploding.

It seems, then, that there is some difference between disconnection and double prevention. Disconnection requires that you start with connection. In the gun case, the sear is originally connected to the firing pin, so that an actual process of removing this connection must occur to trigger the firing. In the case of Billy and Enemy, Enemy was never actually connected to Suzy, so Billy never really disconnected him from Suzy.

Let us summarize the results of this section. First, we have concluded that mechanical systems as wholes have productive capacities, and that sometimes the productive capacities of these systems come from the operation of disconnecting processes among the parts. Productive capacities are not explained in isolation in terms of some kind of physical persistence and transference at the level of individual entities, but instead arise as the result of the organized activities and interactions of entities within mechanisms. On the other hand, in arguing that production does not always need physical connection, we have not thereby collapsed the distinction between two concepts of cause, with which we began. We still find that some things may make a difference to the occurrence of an event without being productively connected to them. Such it would appear is the case for omissions and preventions.

7.8 Causal Relevance and Actual Difference-Making

My emphasis in this chapter has been on production, and I have characterized causal relevance principally in contrast to production. In this section, I would like to say something more directly about the causal relevance relation. Any account of causal relevance must start by answering the question—relevant to what? In general, the answer is *a phenomenon*: some kind of behavior or pattern for which mechanisms are responsible. As discussed in Section 5.1, there are many different types of phenomena, including the regular or irregular behavior of single mechanisms or of types of mechanisms, dispositions, equilibria, and so forth. But, for the moment, our emphasis is on causal relations between events. In this case, relevance is a relationship between the event and the various features of the mechanisms, its entities, activities, and organization that are productive of the event, together with features of the background conditions that are required for the mechanism to produce the event. Relevance is what is captured in the "in virtue of" clause of our canonical form of singular causal claims (Section 7.2).

Since event descriptions come in varying degrees of abstraction, whether some feature is relevant to the production of an event depends upon how abstractly the event is characterized. Consider again Maria cooking the steak. The maximally concrete event of her cooking the steak is a highly particular thing. It is an activity that chemically transforms and rearranges the tissues and their cellular and molecular constituents in some very precise way and countless features of the entities and activities involved will matter to the exact outcome. Just where precisely did that grain of salt land on the surface of the steak? What is the impact of the slight irregularity in the burner element on the heat distribution in the pan? And so on . . .

In practice, however, our interests, even in single-event explanation, focus on only some and not all features of a maximally concrete event. We will characterize the concrete event abstractly in terms of certain of its features—its internal and external temperature, its color, its taste profile, etc. Given an event so specified (explicitly or implicitly), fewer and more abstract features will be relevant to the occurrence of the event. For instance, the amount of salt and the temperature will matter, but not the exact geometry of the salt crystals or the micro-variations in temperature that occur on very small timescales.

Our account of causal relevance has so far showed that there are ways for a mechanistic account that treats causal production as an intrinsic relation to make sense of causal relevance claims. But one question remains: Are these causal relevance claims essentially comparative? This is the position Woodward takes on relevance (or difference-making):

[Difference-making] causal claims involve a comparison of some kind between what happens in a situation in which a cause is present and alternative situations (which may be actual or merely possible) in which the cause is absent or different.   (Woodward 2011, 411)

If it turns out that this is the only way to explicate causal relevance, then one will be left either with a pluralism that says that there are two fundamentally different kinds of causation or with a monism that suggests that apparently intrinsic production relations are in fact explicable in comparative terms.

I think, however, that there is a way out that honors singularist intuitions that causation is an intrinsic productive relation while at the same time making sense of causal relevance. Briefly put, rather than understanding causal relevance as implicitly or explicitly involving a comparison between one case and another, causal relevance relations depend upon abstract features of the single case. When we say that one event causes another in virtue of some relevant feature p, p is an abstract feature of the situation; one that may invite contrasts, but which is not defined by them.

My proposal to connect causal relevance with abstraction is similar in a number of respects to the approach suggested by Michael Strevens, who characterizes the relation between production and relevance in this way:

There are in our causal conception of the world two different kinds of causal relations…But …causal pluralism does not follow. The two causal relations of influence [production] and difference-making [relevance] do not cross-classify causal reality, but rather correspond to concrete and abstract descriptions, respectively, of a single reality, the web of causal influence.
(Strevens 2013, 319)

To be sure, there are substantial differences between Strevens's and my account of this web of influence or production. For Strevens, the web is part of a fundamental level theory of causation, while I have offered an account of production that is multi-layered. Where there are entities engaging in activities, there can be production, even if these entities do not fall within the domain of fundamental physics. And while I agree with Strevens that the productivity of higher-level entities in some sense comes from below, that does not make them any the less productive. But the major point of agreement is that while productivity is concerned with concrete entities and activities, relevance is concerned with abstract difference-making features of those entities and activities. Strevens's account of difference-making diverges from Woodward's in characterizing difference-making in terms of abstract features of the actual case.

Consider the following simple example: Franz returns from the hunt, bumping his head on the door to the kitchen in virtue of its five foot eight inch height. Here we have one event, the walking, producing another event, the bumping, in virtue of an abstract difference-making feature—namely the door's height. The height is a feature of the actual door and it made an actual difference to the actual event. As Franz rubs his head he may muse counterfactually and say that had the door been higher he would not have bumped his head, but the comparison is not essential. The fact that the door height actually made a difference grounds the counterfactual, rather than the other way around.

Actual single-case difference-makers need to be distinguished not only from counterfactual difference-makers but from another sort of actual difference-maker that has been described by Ken Waters (2007). Waters's actual difference-making is a population-level concept. To say that a feature A is a difference-maker for B in a population is to say that what accounts for the contrast between the B and not-B partitions of the population is that the members of the B partition had A, while the not-B members did not. Call such a cause a population-level difference-maker. As an example, consider claims about so-called "single-gene disorders" like cystic fibrosis and Tay-Sachs disease. To say that these are caused by a single gene is not of course to say that the pathology is produced solely by the action of that one gene. Those pathologies involve complicated physiological processes that depend upon the expression of genes and also on countless non-genetic causes. The point of such single-gene claims, when they are true, is that within the background conditions of a particular population, the presence or absence marks the part of the population that has the condition from the part that does not. Population-level difference-makers like these single genes will be actual single-case difference-makers in each of the individuals that has the condition, since the presence of the gene is actually relevant to the occurrence of the condition; but there will in addition be countless other features of the individuals that will make a difference to the occurrence of the condition, but which, because they are shared across the population, will not be population-level difference-makers.

Not all features or circumstances that we intuitively might think of as relevant can be understood as actual difference-makers. Consider again Hall's example of double-prevention. Billy shoots down Enemy, thus preventing him from preventing Suzy from bombing the target. Billy made a difference, but this difference must be understood in a counterfactual sense. In saying that Billy made a difference, we are contrasting the actual world in which Suzy makes it to the target with a non-actual world in which Enemy shoots her down, and asserting that the difference between this world and that comes from Billy's action. In the actual world, neither Billy nor Enemy play any role in the production of the bombing, and so they do not make an actual difference—their activities and interactions are not features of the actual mechanism. This again is the way in which the Billy-Enemy-Suzy case differs from the firing of the gun, where pulling the trigger is part of the actual mechanism of firing, and hence actually makes a difference to the firing.

How exactly do we distinguish between actual and counterfactual difference-makers? If one event causes another, then we know that there is a mechanism connecting them, and this mechanism will have entities whose activities and interactions ultimately produce the event. Actual difference-makers will be features of the actual entities and their activities upon which outcome depends. For instance, Suzy's having fuel in her bomber made an actual difference, because the fuel was burned by her bomber to get to the target. In contrast, neither Billy nor Enemy interacted with any of the entities involved in the production of the bombing.

One way to think about this distinction is epistemically. While we know that the fuel made a difference, because it was actually burned to propel Suzy's jet, we do not know for certain whether Billy's shooting down Enemy would have made a difference, because we do not know for sure whether Enemy would have shot down Suzy had Billy not shot him down—maybe his guns would have jammed, or maybe he would have defected. Who is to say, as none of this actually happened? In terms of omissions and preventions, the only thing we can say *actually* made a difference is that *nothing* prevented Suzy from getting to the target, because actually nothing did.[19] In cases like this, we can still award some credit to Billy, but only in the quasi-causal way discussed in Section 7.7.

---

[19] For a similar account of negative conditions, see Strevens's account of the "one big negative condition" (2008, 200–5).

7.9   Causal Relevance and Inter-Level Causation

Causal relations and dependencies are often characterized as occurring at and across levels.[20] There are high and low-level causes, and causal relationships that seem to cross levels, either top-down or bottom-up. This inter-level causal talk is ubiquitous and seemingly important, but it is difficult to pin it down or determine its metaphysical import. This is because the term "level" is a metaphor (Noble 2013), and is used in many different ways across the sciences (Craver 2007, ch. 5; Love 2011). In this section, I would like to say something about how to make sense claims of higher-level and inter-level causation from the perspective of New Mechanist's ontology and the account of causation I have developed.

The central notion of levels that is implicit in the account of mechanisms and their organization are levels of composition, and the vertical dimension of mechanistic organization I have called mechanistic constitution is a compositional relation. Mechanisms have parts, and their behavior as a whole is constituted by the behavior of their parts. I will follow Craver (Craver and Bechtel 2007; Povich and Craver 2018) in calling this particular species of compositional relation a relation between higher and lower mechanistic levels.

While there are challenges to determining when an entity is a part of a mechanism or where to draw the boundaries between mechanisms (or entities or activities) and their environment (see Section 2.5), if these challenges can be met, levels of mechanisms give an ontologically perspicuous sense of level. As Love (2011) points out, these kinds of compositional levels are local. A particular kind of entity or system can have hierarchical levels of organization, and we can distinguish properties of mechanisms as wholes from properties of their parts. There are no global levels that exist apart from these localized hierarchies, so there is no way to locate an entity, activity, property, or event at some unique and universal level. Sometimes we do use the term "level" to pick out a more global category of things, as when we speak of the psychological level or the neurological level, but while these characterizations can have considerable practical value (Glennan 2015a), they do not pick out ontologically coherent categories.

Craver and Bechtel have argued that talk of top-down (or bottom-up) causation does not imply there really are top-down causes. Instead, they argue that putative cases of top-down and bottom-up causation are examples of mechanistically mediated effects, which they define to be "hybrids of constitutive and causal relations in a mechanism, where the constitutive relations are interlevel, and the causal relations are exclusively intralevel" (Craver and Bechtel 2007, 547). To cite the example of the causes of muscle contraction introduced in Section 2.7, it is not strictly proper to speak of the sliding of filaments within sarcomeres causing the contraction of muscles, from bottom-up. What is really happening is that the sliding of filaments is an interaction

[20] For a variety of perspectives from philosophers and scientists from a variety of disciplines see the special issue of *Interface Focus* on top-down causation (by George F. R. Ellis, Denis Noble, and Timothy O'Connor), vol. 2, issue 1 (2012).

between those filaments, and that the contraction of the muscle is constituted by the sliding of these filaments—involving several levels of mechanistic constitution, from sarcomeres, to myofibrils, to myocytes, and on up to muscles. As another example, consider the searing of the steak, wherein the changes to the steak produced by the searing must be understood as being constituted by the activities of and interactions between the constituents of the steak.

Craver and Bechtel's account seems to me to be fundamentally correct, but I think that the distinction between productivity and relevance can be helpful in understanding the purport of claims that certain kinds of phenomena involve top-down or bottom-up causes. Their claim that causal connections are intra-level should be understood as a claim about production. When two or more entities interact, this produces changes in those entities. As parts of a mechanism, these entities are, by definition, at the same mechanistic level. But when we say that certain causes act at higher or lower levels, we may be saying something about which properties of these entities (or of their environment) are relevant to the outcome.

Suppose, for instance, that Franz is suffering from depression; his doctor prescribes him a set of pills, and he begins to feel relief from his depressive symptoms. The relief might have nothing to do with the pills—but let us suppose that it did. Taking the pills somehow produced a change in his mood. The question, though, is by what mechanism it did so. Here are two possible ways: First, it may be because of the chemical structure of the active ingredients in the pill. Franz might for instance be taking an SSRI—a kind of medication whose net effect is to increase the amount of the neurotransmitter serotonin, and thereby to enhance connectivity between neurons. This change to brain activity would (in some way not really known) constitute the change to mood. The second way the pill might work would be as a placebo. In this case, a set of beliefs that Franz forms about the efficacy of the pill to alter his mood. How this mechanism works is very far from known, but it clearly must be mediated through a complicated set of cognitive states and processes—states and processes that are very much connected with the broader socio-cultural context in which Franz is embedded. Franz's confidence in his doctor is clearly key, and how he comes to have that confidence is no simple story.[21]

In each of these scenarios, the productive causal claim that Franz's taking the pills changed his mood is true, but we get two different canonical causal claims:

Franz's taking the pills changed his mood in virtue of the pills' active pharmacological ingredient.

Franz's taking the pills changed his mood in virtue of a placebo effect.

Intuitively we might think of the placebo case as an instance of higher-level causation, and indeed of top-down causation, because the causally relevant properties are

---

[21] For the sake of simplicity I am telling this as an either/or story, but nothing excludes the possibility that an efficacious medical treatment derives its benefit both from pharmacological and from psychological causes. Indeed, that is quite likely.

psychological and social, which we are inclined to think of as higher-level properties, and mood changes are (we will suppose) somehow constituted by neurological features of the brain, which are lower-level.

These levels are not strictly comparable using the apparatus of mechanistic levels, because the pharmacological and placebo mechanisms are different mechanisms. What is clear, though, is that the causally relevant properties of the pharmacological mechanism are properties of the pill, while the causally relevant properties of the placebo mechanism involve a much larger set of mechanisms for belief formation of which the pill is only a small part.

Let me discuss one further case that we might take to be a case of top-down causation, the phenomenon of frequency-dependent selection. Frequency-dependent selection is a form of natural selection in which the fitness of a phenotype depends upon its frequency in the population (Ayala and Campbell 1974). There are a variety of different biological traits which exhibit frequency dependence; a likely class of examples involve predation. According to what is sometimes called the searching-image hypothesis, predator species identify their prey by a stereotypic search image—that is, an image of what the typical member of the prey species looks like. So, for instance, if the most common form of prey has black coloring, the predator will search based on an image of a black prey. The upshot of this is that prey are less likely to be eaten if they have unusual forms, so prey with unusual forms will be more fit. But to say something about the frequency of a form is not to name a property of the individual animal, but of the population of which it is part.

If the fitness of an individual depends upon a property of the population then we have a prima facie case of top-down causation. Elsewhere (Glennan 2009) I have argued that the proper way to understand this is as a claim about causal relevance rather than causal productivity. As I read it, the various interactions between individual predators and prey are the productive causal interactions that permit or prevent future reproductive success for particular organisms. The relative frequency of the trait does make a difference to the probabilities of surviving predation in particular cases, and a causal explanation of the reproductive success of rare forms will depend upon it, but it is not a productive cause, since productive causes involve events where entities interact.[22]

Much more could be said about the various forms of inter-level dependencies that exist in hierarchies of mechanisms, but the central point is that there is nothing unusual or metaphysically earthshaking in these dependencies. The New Mechanist framework for analyzing constitution and causation provides the resources for understanding how such dependencies are causally and constitutively explained.

---

[22] Roberta Millstein (2013) has taken issue with my analysis of this case, arguing that natural selection involves productive causal interactions between populations. Her critique hinges on the fact that she believes that populations are individual entities. If her analysis is correct, then the interactions between predator and prey populations would be instances of what I have called constitutive production.

7.10 Just Enough Reduction

The New Mechanist approach to causation, with its emphasis on mechanism-dependent production, has a reductionist flavor. The productive capacities of wholes depend upon the capacities of parts. But in many circles, reduction is a dirty word. It bespeaks an ideology of nothing-but, where the whole world is nothing but the sum of the parts. In concluding this chapter, I will argue that the New Mechanism's reductionism is very far from nothing-but reductionism. While it embraces the (to me at least) obvious truth that all things are physical and that their causal capacities must depend upon their basic physical constituents, it also shows clearly why nothing-but reductionism is profoundly mistaken.

The scientific study of food and nutrition provides many excellent examples of reduction gone too far. The journalist Michael Pollan, who is something of a prophet for the whole foods movement, has popularized the term "Nutritionism" to refer to a reductive ideology that has come to dominate the science, marketing, journalism, and politics that surrounds the production and consumption of food.[23] Nutritionism at its core is the idea that foods are simply collections of nutrients—invisible molecular entities that ultimately are responsible for meeting the dietary needs of humans, and other animals as well. Nutritional analysis has its origins in the nineteenth century with the identification of fats, carbohydrates, and proteins as the major constituents of foods. Its cause was advanced by the finer-grained analysis of food, which led to the identification of essential micronutrients—vitamins and minerals. If foods are simply collections of nutrients, then finding a proper diet is a matter of consuming appropriate quantities of various nutrients. Consumption of too much or too little of various nutrients can cause disease and lead to a variety of problems in growth and development.

There is truth in Nutritionism. Foods are, like everything else, made of molecules, and the various kinds of molecular constituents of food play different and essential roles in metabolism. Fats, carbohydrates, and proteins all are metabolized to produce energy, which supports cellular and organismic function. Foods also contain nutrients required for the synthesis of biomolecules used within cells. Proper bodily functions do require these macro- and micronutrients, and failure to ingest them will cause disease and even death. A diet with insufficient iron will lead to anemia; a diet with insufficient vitamin C will lead to scurvy; and so on.

At the same time, the rash of conflicting studies and dietary advice suggest something is deeply amiss with Nutritionism. Fat was bad, but now it is good. Once margarine was the good fat, because animal fat was bad, but now butter is back. Nutritionists and gurus offer conflicting if scientific-sounding accounts of why we should have hi-carb diets, high-protein diets, high-fat diets, and many variations besides.

---

[23] Pollan (2009) is a book-length critique of Nutritionism. Pollan borrows the term from the sociologist Gyorgy Scrinis.

The root of Nutritionism's problem is its nothing-butism. Pollan quotes NYU nutritionist Marion Nestle, who says that "the problem with nutrient-by-nutrient nutrition science is that it takes the nutrient out of the context of food, the food out of the context of the diet, and the diet out of the context of the lifestyle" (2009, 62). One of Pollan's examples nicely illustrates the point. Epidemiological studies provide clear evidence that diets rich in fresh fruits and vegetables reduce the risk of cancer. A plausible nutrient-based explanation is that fruits and vegetables contain antioxidants. Antioxidants can absorb free radicals, which are capable of damaging DNA in ways that can trigger cancers. But these benefits do not appear to be duplicated if the antioxidants are ingested as dietary supplements, and in the case of beta-carotene, one study has suggested that dietary supplements may actually increase the risk. Pollan asks:

What is going on here? We do not know. Maybe it could be the vagaries of human digestion. Maybe the fiber (or some other component) in a carrot protects the antioxidant molecule from destruction by stomach acids early in the digestive process. Or we could have the wrong antioxidant....Or maybe beta-carotene works as an antioxidant only in concert with some other plant chemical or process.   (64)

And so far we have considered only the nutrient's context within the food. But foods are not ingested alone, but in combination with other foods. Is red wine good for you? Epidemiological studies suggest it can be, but whether it is so surely will depend not only upon how much you drink, and on much else besides. It will depend upon whether you are drinking at meals, and perhaps what is being served at meals, and when those meals are served, and what you do before and after them.

All of this dependence on context, on organization, and on nesting of systems and processes within other systems and processes fully comports with the New Mechanism's understanding of the nature of causal processes and systems. For while that account does indeed emphasize the way in which productive capacities of wholes depend upon their parts, it equally emphasizes the central role of organization in systems and processes. So, for instance, it is entirely to be expected on the New Mechanistic account that the way in which a set of nutrients will interact with an animal's digestive system will depend upon the way these nutrients are organized within foodstuffs, as well as the timing of ingestion of these foodstuffs with other activities of the animal.

Nutritionism assumes that foods are, to use a term from Bill Wimsatt (2000), aggregative. An aggregative system is one whose properties are invariant under reorganization. There are some aggregative features both of food and of eaters that will remain constant even if you grind the food or the eater up in a blender—but most of the properties relevant to explaining the role of food and diet in human life and health will not. Emergent properties are for Wimsatt simply the non-aggregative ones. Since the organization of mechanisms will almost always be non-aggregative, the mechanistic account of causation, ontologically reductive as it is, is compatible with and indeed demands at least this form of non-reductive emergence.

Nutritionism also illustrates the craving for generality and its dangers. For understandable reasons, nutritionists, doctors, and the public at large are very interested in generalizations about what constitutes a healthy diet. The experts are happy to oblige with lots of advice, often conflicting: eat a diet low in fats, or at least saturated fats; make sure to eat a minimum daily allowance of vitamins and minerals; avoid refined sugars; make sure you get your omega 3 fatty acids. Part of the reason so much of this advice can be wrong or misleading is that it flows out of errors of Nutritionist ideology: it counsels consumption of nutrients without regard to context.

But just as much of the challenge lies within nature itself. Both eaters and what they eat are complex particulars, and these particulars are heterogeneous. Nutritional generalizations are mechanism-dependent, and variation in composition and structure of individual bits of food and those who eat them will be such that any generalizations will be highly idealized and, for some individuals, outright false. Here are some examples: genetic variations in humans will impact the efficiency with which they can metabolize sugars or fats, so that the amount of calories gleaned from the same meal will differ among individuals. Some individuals will not tolerate lactose; others have difficulties with glutens. And while genetics is certainly responsible for a good deal of the variation in how individuals respond to foods, epigenetic and environmental factors play major roles as well. For instance, much of the work of the gut is accomplished by a vast array of bacteria and other organisms that constitute the microbiome, and variation in microbiomes can account for differential capacities to metabolize foods. We can find similar variations in the nutritional character of foods. The nutritional value of a piece of fruit will depend upon its genetics, the soil in which it was grown, when it was picked, how it was stored, and how it was prepared.

All of this variation and the resulting difficulty with forming true and precise generalizations about nutrition is to be expected from the point of view of the New Mechanism. Far from leading to nothing-but reductionism, its account of causation makes clear just what is wrong with such doctrines. And while it is the case that the productive capacities of foods do depend in part upon the micronutrients of which they are composed, the New Mechanist account of causation can also explain why many features other than the concentrations of those nutrients will be causally relevant to health outcomes.

7.11 Conclusion

In this chapter, I have fleshed out the mechanistic theory of causation by giving an account of the notions of causal production and relevance. We have seen that there is a continuity with the process-based mechanistic accounts of Salmon and Dowe, but that the causal pluralism associated with the activities-based conception of production changes the Salmon/Dowe account in such a way as to allow it to avoid the legitimate objections to it that have been raised by advocates of difference-making approaches to causation. The account of causal production also does not assume anything about the microphysical structure of the world that is inconsistent with our best current scientific theory. While physical theories will almost certainly continue to change, there is little doubt that our account of the microphysical world will continue to exhibit non-classical features, so it is good that we have been able to offer an account of the emergence of classicality, which helps us establish the point where it becomes possible to talk of mechanistic causality.

The account of causation given here, together with the account of mechanistic constitution, provide the resources to make sense of higher-level causal claims, as well as of claims about top-down and bottom-up causation. They also allow us to see how, while the behaviors of whole systems may be grounded in the behavior of their parts, we must abandon simple-minded reductionisms that suggest that wholes are nothing but the sum of their parts, that properties of wholes cannot make a difference to the behavior of the system's parts, or that causation is nothing but causation at some imagined micro-level.

In the final chapter of this book, we will return to the question of explanation, and make use of this account of causation as we sort out different ways—causal and non-causal—that scientists may explain how things in this world hang together.

8

Explanation
Mechanistic and Otherwise

Explanation is, to use Wilfred Sellars's phrase, about showing how things hang together. It is to show what depends upon what. The ontological position I have defended in this book suggests that natural phenomena depend upon mechanisms, so it might seem that this would commit me to a thesis about explanation to the effect that all genuine explanation of such phenomena is mechanistic; but that is not the case. I shall argue that a unified mechanistic ontology is compatible with an explanatory pluralism that recognizes many varieties of explanation that are non-mechanistic and non-causal.

I will begin this chapter with a very general account of what explanation is, arguing that it involves the construction of models that exhibit dependency relations and that unify classes of phenomena. I shall then explore the relationship between this account and Salmon's much-discussed distinction between ontic, epistemic, and modal conceptions of explanation, arguing that Salmon's three conceptions should not be understood as competing accounts of explanation, but instead represent three aspects of any genuine explanation. The remainder of the chapter will be devoted to exploring three kinds of explanation: bare causal explanations, mechanistic explanations, and non-causal explanations.

8.1 Explanation: The Big Picture

The word "explain" is used in many ways. Sometimes we talk of explanation as an activity that human beings, including scientists, engage in; at other times, we talk of one thing in the world explaining another; and at others we speak of theories or models explaining phenomena. Craver has nicely summarized the various ways we speak of explanation by identifying four modes, which he characterizes with the following examples:

1. Jon explains the action potential (Communicative Mode)...
2. The flux of sodium (Na+) and potassium (K+) ions across the neuronal membrane explains the action potential. (Ontic Mode)...
3. The Hodgkin-Huxley (HH) model explains the action potential. (Textual Mode)...
4. Jon's mental representation of the mechanism of the action potential explains the action potential (Communicative Mode). (Craver 2013a, 30–3)

The first three of these modes clearly represent recognizable ways in which the term "explain" is used, and the fourth, while forced, is meant by Craver to express the sense in which explanations are things that cognizers can grasp, and that produce understanding.

While all these modes are legitimate, my starting point for a general account of explanation will be the textual mode. Explanation is a matter of *representing* what depends upon what, and for that, one needs a representational vehicle, or, as I prefer to say, a model. Given Giere's (2004) account of models and representation (see Chapter 3), we can see that explanation is one of the purposes for which a model can be used:

S uses X to represent W for *explanatory* purposes P

Models can be used for many non-explanatory purposes—e.g., description, prediction, and control—but explanation is a key purpose of models in science. Also, it should not be supposed that there is just one explanatory purpose with respect to a given target. One may wish to explain different aspects of the phenomenon at different grains and in contrast to different alternatives. The plurality of purposes and models is what accounts for the fact that there are many kinds of explanations that can be offered of some event or other phenomenon.

So what makes a model explanatory? I offer two entirely non-original suggestions. The first is **dependence:**

> Explanatory models must show that something (the explanandum) depends upon something else (the explanans).

Dependence is, I think, a sine qua non for explanation, but it is not the only virtue explanations can display. A second is **unification:**

> Explanatory models may show that two or more things are similar, either in what they depend upon, or in what depends upon them.

These two features seem to be characteristic of explanation as such.[1] They are exhibited by, but not restricted to, explanations of natural phenomena. Mathematical explanations, moral explanations, and aesthetic explanations all seem to exhibit them. Since my topic is the explanation of natural phenomena—natural explanation, as I will henceforward call it—I shall not try to argue for or illustrate claims about these other domains. Nonetheless, it is helpful to observe that these norms apply to explanations of things that are not part of nature.

Dependence comes in many forms. Much of natural explanation is causal explanation, and causation is a kind of dependence, as effects depend upon their causes. There are also the many forms of mechanism-dependence that have been discussed throughout this book. Mechanism-dependent phenomena depend upon the organized activities

---

[1] The idea that causal explanation involves showing dependencies is common to a great variety of forms of causal and mechanical explanation (e.g., Salmon 1984; Woodward 2003; Strevens 2008). Classic statements of the unificationist approach include Friedman (1974) and Kitcher (1989).

and interactions of the mechanism's parts. This kind of dependence involves both causal and constitutive relations. Finally, there are, if recent work in the philosophy of science is on the right track, many explanations of natural phenomena that are not mechanistic or even causal. I will discuss some of these forms of non-causal explanation in the second half of this chapter, arguing that, even though such explanations do not show how phenomena depend upon causes or mechanisms, they are explanatory precisely because they exhibit relations of dependence.

Formally we can think of dependence as a class of relations that is irreflexive (A does not depend upon A), non-symmetric (if A depends upon B, B may or may not depend upon A), and transitive (if A depends upon B and B depends upon C, then A depends upon C). These formal similarities are analytic consequences of our concept of dependence, but I do not think they require us to think of dependence as one thing. It is better to think of a family of dependence relations bound together by shared formal properties.[2]

While dependence is the sine qua non of explanation, we should understand this requirement broadly enough that information about *independence* can be explanatory as well. Assuming (as I do) that explanation is not an all-or-nothing affair, identifying features of a situation that are independent of the explanandum can often be explanatory. Equilibrium explanations (to be discussed below) operate in part by showing that the end state of a system is largely independent of its initial state. To know this is to understand something important about the system. Similarly, explanations of robustness or resiliency, which show how a system's state is insensitive to various kinds of interventions or perturbations, contain explanatory information (Wimsatt 2007; Gross 2015). Discerning that some phenomenon is independent of an event, fact, or process is especially explanatorily salient in circumstances in which that event, fact, or process was prima facie likely to have been relevant. If in the airplane crash we have ruled out mechanical failure, we have ipso facto shown that the airplane crash depends upon something else. And that is, while nothing like a full explanation, explanatorily relevant information.

While some advocates of explanatory unification models have seen unification as a sine qua non of explanation, I am taking it to be a desirable but non-essential feature. Unifying explanations lump things together, as when Newton showed that celestial and terrestrial mechanics depend upon the same laws and principles. But explanatory progress can also be made by showing how some apparently similar kinds of phenomena

---

[2] My formulation allows for cycles of interdependence (A depends upon B which depends upon A). This seems to me essential because so many systems whose behavior we seek to explain involve feedback and interdependencies between parts and wholes. My formulation is more permissive than typical formulations of another much-discussed kind of dependence relation—grounding (Schaffer 2009a; Fine 2012). Advocates of grounding typically see it as an explanatory relation (or family of relations), but they usually stipulate that grounding relations are asymmetric rather than non-symmetric and that they require least elements (i.e., a set of facts or things that ground everything else). I shall not delve into these issues here, but I do think that the idea that grounding relations are explanatory (or at least back explanations) is illustrative of the general principle that explanatory relations are dependence relations.

depend upon different things (e.g., different causes or mechanisms)—the strategy often known as splitting.[3] And equally it is possible to offer explanations of singular phenomena—by showing what the phenomenon depends upon—without regard to whether the explanatory dependence fits within some broader unifying pattern. Much of natural and human historical explanation, for instance, appears to have this character.

Similar phenomena often have similar causes that depend upon similar mechanisms and conditions. What explains the blue appearance of Franz and Sisi's eyes? The phenomenon turns out not to be due to blue pigmentation, but to a kind of scattering of light called Tyndall scattering. It is analogous to the Rayleigh scattering mechanism that gives the sky its blue appearance. In both cases light traveling through a transparent medium (the sky or the stroma of the iris) is backscattered in shorter blue wavelengths. The blueness not just of Franz and Sisi's eyes, but of the sky above them, is caused by a similar kind of scattering mechanism. That's unification.

The reflective capacity of Franz and Sisi's eyes is not the only explanatory target we could consider. We could also seek to explain why Franz and Sisi and their blue-eyed brethren have this capacity. And we can consider such a question from the point of view both of proximal and of ultimate causes (Mayr 1961). For instance, we can also take the explanandum to be the structural features of the eye that give rise to this reflective capacity, and describe the developmental mechanisms that are responsible for the construction of the iris, with its distribution of pigment. Certainly, to the extent that eye color is largely determined by a small set of genes, we can show how blues eyes depend upon those genes. And the fact that Franz and Sisi have similar genes controlling their eye color may in turn be explained by their ancestry (e.g., by the fact that they share many relatives, or that they come from populations in which certain genes predominate, as a result of selection or drift). Put generally, it appears that the explanation of human eye color is unifying because the mechanisms (evolutionary, genetic, developmental, and optical) that account for certain eye colors are similar across the class of humans.

In the eye color example, what unifies a class of phenomena as an explanatory target are common causes and mechanisms responsible for the existence and characteristics of these phenomena. But explanatory dependences can run the other way, in the sense that a class of phenomena can be unified not by what they depend upon, but what depends upon them. As an example of this, consider how we explain the rules of card games. Suppose I am explaining the rules of Euchre to someone, and I tell them that trumps are determined by a certain bidding process (we can spare the details), and that the trumps consist of all cards of a selected suit plus the Jack of the same color (e.g., if the trump suit is spades, the Jack of clubs is trumps). The Jack of the trump suit

---

[3] Lumping and splitting are terms that have a long history in biological taxonomy. Craver (2004; 2009) has explored how these strategies are connected to mechanistic conceptions of natural kinds. See also Section 4.1.

is the high trump, followed by the other Jack of the same color, followed by Ace, King, Queen, 10, and 9.

In calling a suit or collection of cards trumps, I am defining the collection by identifying a functional role of these cards within the game. Trumps are what they are because of what they do. In Euchre and most other cards with trumps, cards are played (one from each player) in tricks, with the highest card of the led suit winning the trick, unless someone plays trumps, in which case the highest trump card wins. This is what it means to be trumps. While we speak of a trump suit, in the case of Euchre, a Jack from another suit becomes part of that suit. It is not the symbol on the card, but the role it plays that defines a card as trumps.

To identify a set of cards as trumps is to identify explanatory dependencies. We show that the fact that a certain player wins a certain trick depends upon their playing the highest trump card if trump cards are played. But the kind of explanatory work done by characterizing cards as trumps is also unifying. For one thing, in characterizing sets of cards as trumps, we are explaining not just a single round of play, but a whole class—all the tricks played in the round, as well as in all other rounds. Equally important, the explanation is unifying because it places a particular game like Euchre within a whole class of games that are trick-based and utilize trump suits—for instance, Bridge, Whist, Oh Hell, and Skat. These games determine trumps in different ways, but in all of these, trump cards have a similar effect upon play, so it enhances understanding of a game to see the relation of its rules to rules of other games. But again, while this is certainly an explanatory virtue, it is by no means necessary. Even if there were but one card game using trumps, to understand how the outcome of a game depended upon trumps would be sufficient to provide genuine understanding.[4]

Explanatory unification is connected to problems of generality and mechanism kinds that have been discussed in earlier chapters. In Chapters 3 and 4, I argued that the right way to understand kinds and generality was through the models-first principle. To classify a collection of mechanisms as a kind is just to identify a common model that can be used to describe all of the members of that collection. So, for instance, we understand (human) blue eyes as a kind in part because the same model can be used to describe the parts, organization, activities, and dispositions of all tokens of human irises that appear as blue. In such a case, the work of explanatory unification is done by the shared model.[5]

---

[4] Rule-based explanations provide a good example of a class of dependency explanations that are not, straightforwardly at least, causal. Rules of card games explain, but rules are not like causal generalizations, in the sense that they depend upon intentional agents to follow them, and the rules can of course be (accidentally or deliberately) broken.

[5] My account of modeling, and with it of explanatory dependence and unification, embraces what Batterman and Rice (2014) have called a "common features" account of explanation. Models explain because they share common features with their targets, and models unify because of common features shared across multiple targets. Though Batterman and Rice have criticized such accounts, I concur with recent responses (Lange 2015; Povich 2016) that question the cogency of these criticisms.

But while sharing an explanatory model is sufficient for explanatory unification across targets, a same-model requirement is doubtless too strict. Explanatory unification can be achieved when one shows how the same kind of model (i.e., models constructed using similar principles) can be used to explain otherwise diverse phenomena. Consider the unification of celestial and terrestrial phenomena. One of Newton's achievements was to show that the motions of the earth, the other planets, and the moons depend upon the same gravitational forces that govern projectile motion and motion of falling bodies. It would be quite a stretch to say these phenomena are explained by the same model, but the models all use the same principles, embodied in Newton's laws of motion and in the law of universal gravitation, and the shared use of these principles brings about explanatory unification.[6]

More generally, the analysis of types of mechanisms in Chapter 5 showed that there are many cross-cutting ways to classify mechanisms—for instance, by the types of organization, etiology, or activities involved. These categories are powerful tools for unifying disparate phenomena and yielding understanding. If, for instance, we discover that physical systems and economic markets exhibit similar self-organizing properties, which can be modeled using similar mathematical tools, we do not just have models that show what these phenomena depend upon. We have also achieved some amount of unification between physical and economic domains. Or again, a great many kinds of systems can be modeled as involving agents that may adopt differing strategies to maximize their goals or interests. Such systems, be they political, economic, or biological, can be studied using common models from game theory.

While unifying explanations are explanatory achievements, showing that apparently similar phenomena are dependent upon different causes or mechanisms is likewise an achievement. The strategy of splitting is well illustrated by cases from the field of cancer research. Like many diseases, cancer was originally identified by its symptoms—tumors or growths—to which the word "cancer" is etymologically related. Classification of kinds of cancers begins with identifying the organ or region where the cancer occurs: for instance, breast, lung, or prostate cancer. Cancer research over the last century has led to progressive rethinking of these categories. Cancer came to be thought of as a disease of uncontrolled cell division, with detectable tumors being neither necessary nor sufficient for cancer. Cancers can be split according to the types of cells involved—for instance, carcinomas, sarcomas, and leukemia. These classifications rearrange the taxonomy because cancers can affect similar cell types in different organs (e.g., carcinomas in breast or colon) and the same organ can have cancers originating in different kinds of cells (e.g., sarcomas and carcinomas of the breast). Further taxonomic splitting occurs as the genetic basis of different varieties of cancer becomes better understood.

---

[6] There are not clear rules that allow us to decide whether two models are (qualitatively) identical. The line between changing the parameters in a model and changing models is fuzzy. See discussion in Chapter 3.

The importance of splitting strategies in taxonomic reform suggests the primacy of dependence as an explanatory value. Splitting involves showing that grossly similar phenomena depend upon different mechanisms. Especially if (as in clinical medicine) one's concern is to understand the causes of symptoms in a particular case, splitting is essential, as it allows for more successful and targeted interventions. To know, for instance, that a breast cancer is of a variety that is fed by estrogen and progesterone is both to understand what that cancer depends upon and to suggest an intervention (suppressing production of these hormones) that can fight the cancer.

The explanatory value of splitting is an epistemic consequence of the ontological reality of the particularity of causes and mechanisms. But the fact that finer-grained mechanistic information gives rise to better understanding of particular phenomena does not imply that lumping cannot be an explanatory move. Sometimes coarse-grained models that lump heterogeneous phenomena can provide genuine understanding of broader patterns of phenomena. As a case in point, we can consider another example from cancer research—the multi-stage model of carcinogenesis.[7]

The multi-stage model is thought to explain patterns of incidence of cancer in a wide range of carcinomas. Although cancers show wide variation in their frequency, a significant number of cancers in diverse areas of the body are similar in the exponential growth of incidence with age. In one of the first papers on the multi-stage model, Armitage and Doll describe the pattern:

[T]wo hypotheses about the mechanism of carcinogenesis have been put forward, which have been derived from analysis of cancer mortality statistics. Fisher and Hollomon (1951) used statistics from the United States for cancer of the stomach in women, and Nordling (1953), classing all sites together, used statistics for cancer in men from Britain, France, Norway and the U.S.A. Both found that, within the age group 25–74 years, the logarithm of the death rate increased in direct proportion to the logarithm of the age, but about six times as rapidly; in other words, the death rate increased proportionally with the sixth power of the age. (Armitage and Doll 1954, 1)

The multi-stage model is a simple mathematical model that is meant to account for this pattern. The basic idea is that carcinogenesis is a multi-stage process in which a series of mutations accumulate in cells over time. Armitage and Doll showed that if these mutations were assumed to have a constant chance of occurring through a person's lifetime, then the relationship between age and incidence would follow a power law. In the model, the particular power of age is one less than number of hypothesized mutations—so the fact that these cancers increased with the sixth power of age would follow from a model in which there were seven stages of carcinogenesis.

---

[7] I borrow this example from Anya Plutynski (2013). Plutynski uses the multi-stage model as an example of the phenomenon that has come to be called integrative research. She helpfully draws a distinction between integration as a research strategy and unification as an explanatory ideal. While I am focusing here on the possible unifying power of the model, I do not mean to equate unification with integration.

This model, if it is right (i.e., if it is relevantly similar to the class of target systems to which it is being applied), is an example of what I have earlier called a how-roughly model.[8] It does sketch very roughly some features of the carcinogenesis mechanism, but it is completely lacking in biological detail. What matters is only the number of stages and the constant probability of occurrence of each stage. But this is enough to say something important about the structure of the mechanisms that account for the pattern, and it is also enough to explain why those cancers for which the model assumptions fail to hold do not exhibit this pattern. For example, there are substantial divergences in these patterns in hormonally driven cancers like ovarian or prostate cancer, where changing rates of hormone production in different life stages will undermine the constant probability assumption. Similarly, the pattern fails to hold in cancers where the probabilities of mutations are changed by exposure to environmental insults like cigarette smoke, industrial chemicals, or radiation.

The model achieves unification by identifying some abstract but non-trivial features of an otherwise quite heterogeneous set of target phenomena. Importantly, the unification will only be genuine if these features actually exist in the targets. Characterizing the pattern is not explanatory if it is merely an exercise in curve-fitting, which is why those who have developed the model have rightly exhibited caution about the genuine explanatory value of the model. But even if this model does provide genuine explanatory insight, this is not to say that the finer-grained biologically detailed models are wrong or less good. There are not really competing explanations here, but complementary explananda.

Let me recapitulate the most important points of this section. First, I have argued that explanation is a matter of showing what depends upon what. Explanation involves construction of a model in which dependencies in the model represent dependencies in the explanatory target. Second, knowledge of what does not matter to some phenomenon, i.e., knowledge of what is independent of or irrelevant to the phenomenon, does provide indirect information about what depends upon what. Third, unification is a virtue, but one that is not necessary or possible for all explanations. And when explanations are unified, it is because a class of phenomena can be represented by a common model, or at least by related models, which illustrate how the various phenomena depend upon the same kinds of things. Finally, just as knowledge of independence can give us explanatory insight into dependence, knowledge of when phenomena are not unified (i.e., splitting) can give us explanatory information about what is (not) like what.

8.2   Explanation: Ontic, Epistemic, and Modal

One of Wesley Salmon's many contributions to the philosophy of scientific explanation was his influential account of the "three conceptions" of scientific explanation. Writing

---

[8] Plutynski's review of the history of this model suggests that, while it has been subject to criticism and modification in the more than fifty years since it was proposed, it continues to be considered theoretically valuable.

in the 1980s, Salmon (1984; 1985) sought to situate his own changing views of explanation within the philosophical discussion that began with Hempel and Oppenheim's (1948) deductive-nomological model. Salmon saw his own causal-mechanical account of explanation as exemplifying an ontic approach to scientific explanation, one that differed in important ways from two other conceptions, which he called the epistemic and the modal.

Salmon's conceptualization of the explanation debate has been influential in shaping discussions since that time, and it has in particular surfaced in debates over the nature of mechanistic explanation. Among New Mechanists, the chief defender of the ontic approach has been Carl Craver (2006; 2007; 2013a), who sees his work on mechanistic explanation as extending Salmon's approach. Bechtel and collaborators have in contrast argued for an epistemic account of mechanistic explanation (Bechtel and Abrahamsen 2005; Bechtel 2008; Wright and Bechtel 2007), arguing that explanation always involves representation. For a stark statement of the contrast, consider these two passages, first from Craver:

Conceived ontically...the term explanation refers to an objective portion of the causal structure of the world, to the set of factors that produce, underlie, or are otherwise responsible for a phenomenon. Ontic explanations are not texts; they are full-bodied things. They are not true or false. They are not more or less abstract. They are not more or less complete. They consist in all and only the relevant features of the mechanism in question. There is no question of ontic explanations being "right" or "wrong, " or "good" or "bad." They just are.   (Craver 2013a, 40)

And then from Wright and Bechtel:

Explaining refers to a ratiocinative practice governed by certain norms that cognizers engage in to make the world more intelligible; the non-cognizant world does not itself so engage. One way to appreciate this point is to recognize that mechanisms are active or inactive whether or not anyone appeals to them in an explanation. Their mere existence does not suffice for explanation; the systemic activity of a mechanism may be responsible for the presence of some...phenomenon Φ, but Φ is not explained until a cognizer contributes his or her explanatory labor.   (Wright and Bechtel 2007, 50–1)

Craver suggests that explanations, conceived ontically, are in the world, and that mechanisms themselves explain their phenomena. Wright and Bechtel, on the other hand, think that the ontic mode, taken literally, involves a category mistake. Explanations may refer to things in the world, but they are not things in the world. They are essentially texts (or models) constructed by cognitive agents.

If we take these two passages as characterizing the ontic/epistemic divide, the account I have offered would seem quite clearly to side with the epistemic. I have argued that explanations are models constructed by agents for explanatory purposes, and that they can be more or less complete and more or less abstract. Nonetheless, I think that a fuller and more charitable reading will reveal that the disagreement between Craver, Bechtel, and others is overstated, because charitable readings of each position show that both positions recognize that good explanations must satisfy both ontic and epistemic constraints (Illari 2013).

Advocates of each conception should (and I think do) agree on the several points. First, it is clear, as Craver has suggested, that the word "explain" is sometimes used to refer to a relation between things in the world, and sometimes used to refer to representations of or communications about things in the world. Second, Craver's defense of the ontic conception is not meant to deny that explanations in science use models, and that such models can be more or less complete, more or less abstract, or more or less idealized. Third, Bechtel's defense of the epistemic conception is not meant to deny that explanations, to be successful, must represent things in the world. The disagreement is mostly one of emphasis. Craver argues that questions about the adequacy of an explanation are largely settled by features of the world, rather than features of our models. For example, the height of the flagpole explains the length of the shadow, but not conversely, because the former, but not the latter, relation represents the direction of causal dependence in the world. Bechtel, on the other hand, argues that to grasp an explanation we must have a model, and that to successfully function as an explanation, it must be something that we can understand, and much of his work is concerned with describing the kinds of models that can meet that end.[9]

The recent debate over ontic versus epistemic conceptions has generated more heat than light, and I think that the reason for this largely rests with faults in Salmon's original account of the three conceptions. The confusion arises from the assumption that the ontic, the epistemic, and the modal represent three competing conceptions of explanation, rather than three complementary aspects of explanations. If we treat them as complementary aspects rather than competing conceptions, we can resolve the recent epistemic debate, and, more interestingly, we can integrate the epistemic and ontic with the mostly neglected modal conception.[10]

To see how we might do this, it would be helpful to return to Salmon's original statement of the distinction.[11] Here is the opening passage from the paper Salmon presented at the 1984 meeting of the Philosophy of Science Association:

---

[9] Probably the most substantive disagreement involves differing commitments to scientific realism. Craver is committed to a fairly strong form of scientific realism, because he thinks there must be causal structure "out there" which can be explored and discovered, and that it is this causal structure that provides constraints on what counts as a genuine explanation (Craver 2013, 40–1). Bechtel has recently (personal communication) been moving toward a position he calls "west coast idealism," which emphasizes the inevitable idealization involved in representing the world as consisting of mechanisms. Mechanistic representation involves drawing boundaries between mechanisms and their environment that are not there in nature. While here is not the place to evaluate the nature of their realist commitments, I think it fair to say that Bechtel, notwithstanding his idealism, agrees that there is something "out there" that in part determines the extent to which our explanatory models are adequate.

[10] The modal has not been entirely neglected. Lange (2013), for one, has been trying to revive it.

[11] I am not certain what the earliest statement of this distinction was. Salmon delivered this paper in the same year as his landmark book *Scientific Explanation and the Causal Structure of the World* (1984) was published, and the material on the three conceptions to be found in the book largely overlap the material found in his paper.

When one takes a long look at the concept (or concepts) of scientific explanation, it is possible and plausible to distinguish three fundamental philosophical views. These might be called the epistemic, modal, and ontic. They can be discerned in Aristotle's Posterior Analytics and they are conspicuous in the twentieth-century literature. In its classic form—the inferential version—the epistemic conception takes scientific explanations to be arguments. During the period of almost four decades since the publication of the landmark Hempel-Oppenheim article, "Studies in the Logic of Explanation" (1948), the chairman of this symposium (Carl G. Hempel) has done more than anyone else to articulate, elaborate, and defend this basic conception and the familiar models that give it substance (Hempel 1965), though it has, of course, had many other champions as well. According to the modal conception, scientific explanations do their jobs by showing that what did happen had to happen. Among the recent proponents of this conception, Rom Harré and Edward Madden (1975), D. H. Mellor (1976), and G. H. von Wright (1971) come readily to mind. The ontic conception sees explanations as exhibitions of the ways in which what is to be explained fits into natural patterns or regularities. This view, which has been advocated by Michael Scriven (1975) and Larry Wright (1976), usually takes the patterns and regularities to be causal. It is this third conception—the ontic conception—that I support.   (Salmon 1985, 293)

To understand the motivations for Salmon's three conceptions, it is important to place them in the context of his own interpretation of the development of philosophical thinking on scientific explanation, and his own particular concerns about the challenges facing then existing models of scientific explanation. Salmon identifies the epistemic conception chiefly with Hempel's covering law model of explanation. On that view (at least on Salmon's reading) explanations are arguments, and there is a close connection between explanation and prediction. Salmon rejects Hempel's and other logical empiricist accounts chiefly on the grounds that they could not handle the traditional objections raised to these accounts—e.g., explanatory asymmetries like the flagpole and the shadow and the explanation of low-probability events. Second, the modal conception associates explanation with non-Humean conceptions of necessitation involving causal powers or laws of nature. Salmon rejects the modal conception chiefly because he thinks it cannot be squared with indeterminism. Third, his interpretation of the ontic conception, and his own account of causal-mechanical explanation, has a decidedly Humean flavor—explaining is a matter of fitting what is to be explained into "natural patterns or regularities."[12]

It is striking to see how much the debate has changed since Salmon originally framed these distinctions. Certainly Bechtel's commitment to the epistemic conception is not a commitment to the inferential version. New Mechanists are of one voice in seeing mechanistic explanation as an alternative to covering law conceptions of explanation (Machamer, Darden, and Craver 2000; Glennan 2002a; Bechtel and Abrahamsen 2005;

---

[12] Salmon has a complex and unsettled relationship with Humean and empiricist approaches to causation. Famously he conceded in his 1984 book that he needed a counterfactual criterion for mark transmission to distinguish causal processes from pseudo-processes. But he appears never to have been happy about this, and subsequently renounced his mark-transmission account in favor of Dowe's decidedly more empiricist conserved quantity theory (Salmon 1997). See Section 7.1 for a summary of that history.

Craver 2007). Similarly, Craver's commitment to the ontic conception does not involve an endorsement of a Humean conception of causation. Both the activities views of MDC and the manipulability approach Craver (2007) later adopts are anti-Humean. Finally, supporters of various forms of anti-Humean causal realism have adopted accounts of causal powers or natural necessity that have modal elements but do not assume determinism (Cartwright 1989; Humphreys 1989; Glennan 1997a).

My suggestion is that Salmon's reasons for preferring the ontic over the epistemic and modal conception were not reasons for preferring this conception as such, but rather reasons to reject particular accounts he associated with the epistemic and modal approaches. Perhaps Salmon was right that all three conceptions are traceable to Aristotle, but even if this is so, we need not accuse Aristotle of inconsistency. Perhaps the reason one finds hints of each in Aristotle is that they are not competing conceptions, but rather represent different aspects of successful explanation.

To recognize the epistemic aspect of explanation is to recognize that explanation always requires representation. While it may be true that we sometimes speak of one thing in the world explaining another, or of explanations being out there to be discovered, the act of scientific explanation is always an act of describing or representing those things. Moreover, for an explanation to succeed, it must be understood,[13] and theories of scientific explanation should explore the kinds of representations that are successful in promoting understanding. To recognize the ontic aspect of explanation is to recognize that whether a proffered explanation makes the grade will depend upon what actually occurs in the world. Key questions, like whether a model genuinely explains, rather than simply describes or predicts, will be answered by exploring the causal and structural relationships in the worldly systems and processes that the model describes. Finally, to recognize the modal aspect of explanation is to recognize that explanations are ultimately about dependence. For dependence, which I have suggested is the sine qua non of explanation, is, broadly speaking, a modal relation.

This last point requires some elaboration. As Salmon understood the modal conception, modal explanations involve "showing that what did happen had to happen." The notion of natural necessity appealed to in the phrase "had to happen" is, I would argue, too strong—but the idea that explanations show more than what did happen is just the idea that explanation is distinct from description, and about this there is little doubt. This is in fact the very intuition behind the requirement of dependence. Among the things that happen, some things depend upon other things, and to show what depends upon what is to explain some things by appeal to the things they depend upon. Dependence is not necessitation, because something in the world (e.g., an event or process) may depend upon many things, none of which necessitates it. The idea that some things depend upon others also does not require that determinism is true. It is,

---

[13] Not in a psychological sense, but in a warranted epistemic sense. If you believe in witches, you may feel you understand the causes of your complaints, but you have not grasped that upon which your complaints actually depend.

for instance, consistent with the idea that there are non-deterministic causal powers or single-case propensities.

I shall not here try to offer an account of the kind of modal relationships that scientific explanation requires, though the relationships of causal production and relevance that I have outlined in earlier chapters exemplify one way to understand the modalities involved in causal explanation. Also, as I shall elaborate below, I do not think that causal dependencies, however understood, are the only sorts of dependencies that figure in the explanation of natural phenomena. I would, however, say that to accept the modal aspect of scientific explanation is to believe in real dependence. Explanatory features of the world, whether causal or not, have modal import. That is just what genuine dependence requires.

8.3 Three Kinds of Natural Explanation

All natural phenomena have causes, but that does not mean that all explanations of natural phenomena are causal. To explore why this is so, we will need to get more specific about what counts as a causal explanation and what does not. I will divide natural explanations into three kinds—bare causal, mechanistic, and non-causal—which I briefly characterize as follows:

Bare causal explanations identify events, properties, or states of affairs that either cause or are causally relevant to some phenomenon. Bare causal explanations identify one or more things upon which the explanatory target causally depends, but they do not identify the mechanism in virtue of which this dependence exists.[14] If, for instance, I say of a fire that it was caused by a lightning strike, I have identified an event that contributed to the production of the fire, but I have not described the mechanism by which this contribution occurred. Bare causal explanations tell you (some of) the causes of the explanandum phenomenon, but they do not tell you how those causes contributed to the production of, or otherwise made a difference to, the phenomenon. For this reason, we can call bare causal explanations *what-but-not-how explanations*.

Mechanistic explanations, by contrast, are *how explanations*. They show how some phenomenon or class of phenomena depend upon mechanisms. Mechanistic explanations proceed in both the horizontal and vertical dimensions, showing how the organized activities and interactions of some set of entities cause and constitute the phenomenon to be explained. We have seen many examples of mechanistic explanations throughout this book. Mechanistic explanation is a species of causal explanation, because (even for constitutive mechanistic explanation) it always involves characterizing the activities and interactions of a mechanism's parts.

Non-causal explanations are a heterogeneous lot that I define negatively as those explanations that show varieties of dependence that are not causal or mechanistic. We

---

[14] I take bare causal explanations to include explanations that cite features like absences that are not causally productive, but which are causally relevant to an effect. See Chapter 7.

can think of them as *why-but-not-what-or-how explanations*, because they do not identify particular causes or processes that contribute to the production of some phenomenon.

Each of these varieties of explanation comes in singular and general forms. Both singular events or states of affairs or classes of or patterns within events or states of affairs are susceptible to bare causal, mechanistic, and non-causal explanations.

In what follows, I shall elaborate on each of these categories. Much of the work for bare causal and mechanistic explanation has been done in Chapter 3, so I will here only briefly situate these varieties of explanation within the general framework of natural explanation I have sketched. The bulk of my time will be devoted to exploring some varieties of non-causal explanation that have recently been advanced, with the aim of showing how these differ from but complement bare causal and mechanistic explanations.

8.4 Bare Causal Explanations

Bare causal explanations show what depends upon what without showing why or how this dependence obtains. The causal claims required are established, broadly speaking, by observational and experimental methods like Mill's methods of agreement and difference or controlled experiments. Ontologically speaking, causal dependencies require the existence of mechanisms, but bare causal explanations are silent on what those mechanisms are.

Semmelweis's explanation of the causes of childbed fever in the Vienna General Hospital provides a good example of a bare causal explanation. To briefly recall the story, already recounted in Chapter 6, Semmelweis sought to explain the epidemic of puerperal (or childbed) fever among mothers giving birth at the Vienna General Hospital during the 1840s. His first observation was that the division of the hospital to which the women were admitted appeared to be causally relevant, since the death rate from puerperal fever for women in the First Division was three to four times that of women admitted to the Second Division (6.8–11.4% versus 2.0–2.7%) (Hempel 1965).

There are many facts about this case that we might want to explain. For instance, we might want to explain why in the 1840s the death rate in the Vienna General Hospital rose from previous levels; we might want to explain why death rates in one ward were higher than another; we might want to explain more generally the higher incidence in puerperal fever in hospitals than in home births; or we might want to explain the causes of puerperal fever for specific women who contracted it.

Let us start with the single case. Suppose Clara contracted puerperal fever (call this event *e*). What caused her to contract it? A first explanation might simply be that Clara contracted puerperal fever because she delivered her baby in the First Division (call this *c*). If the claim that *c* caused *e* is true, that is, if there exists a mechanism by which *c* contributes to the production *e*, then that claim provides a bare causal explanation of *e*.

Note that the mere fact that there is a higher incidence of puerperal fever in the First Division is not sufficient to guarantee there is such a mechanism, because it might be the case that that mechanism did not depend upon Clara's being in the First Division. Semmelweis's account suggested that the source of infection for women in the First Division was the bacteria carried from cadavers by the unwashed hands of doctors and medical students who had just performed autopsies, and judging by the effectiveness of requiring hand-washing, that quite likely could have been the cause in Clara's case. But there are other possibilities. For instance, Clara might have acquired the infection just prior to being admitted to the First Division. Suppose, though, that the infection was actually introduced into Clara's body by a medical student who had performed an autopsy. Then the claim is that Clara's admission to the First Division explains her contracting puerperal fever.[15]

One might object that the mere fact that there is a causal connection between her admission to the First Division and her fever is not explanatory. After all, until Semmelweis identified the mechanism of transmission, it was a complete mystery *why* admission to the First Division should be linked to puerperal fever. However, given the basic characterization of what makes something an explanation, this fact is (minimally) explanatory, as it does show something upon which Clara's fever causally depends. One can imagine Clara's distraught family, seeking to understand the causes of her death, discovering that she had been admitted to the deadly First Division. At this point they might say, "So that is why she contracted the fever." And they would, ex hypothesi, be right; they would understand something. Of course, there is much that they would still not understand—specifically they would not understand why and how Clara's fever depended upon the division to which she was admitted. But this is just to say that they have a bare causal explanation, rather than a mechanistic explanation.

It bears repeating that this explanation would only be genuine so long as the under-lying facts about the source of infection were as I have postulated. Indeed, one of the reasons that one should not be satisfied with bare causal explanations is that the methods used to establish them (especially in probabilistic cases like this) are far from certain. There are many alternate scenarios in which being admitted to and treated in the First Division could be correlated with higher rates of fever without being caused by it. For instance, we might imagine that the women admitted to the First Division came from poorer neighborhoods, and that their increased susceptibility to puerperal fever came from differences in diet. Equally important, in the absence of further knowledge of the mechanisms involved, intervention to solve the problem would be difficult.

---

[15] Another more complicated case is if the infection were introduced during her time in the First Division, but through some other source. Suppose anachronistically that Clara's husband was there to cut the cord, and somehow he introduced the infection into Clara's body. The infection occurs in the rooms of the First Division, but its source is from outside of the hospital. Since the causal process of admission to the First Division got Clara into that room, in some sense it appears to contribute to the infection. But intuitively the counterfactuals are different, since her husband might have infected her regardless of where she gave birth.

Short of shutting down the First Division, Semmelweis and his colleagues had no recourse until they figured out what was going on in the First Division that was increasing the incidence of fevers.

One final point we should consider is whether or how bare causal explanations invoke models. I have argued above that explanations always involve the invocation of a model for an explanatory purpose, but it may be thought that there is no model here. When, for instance, I said that admission to the First Division explained the occurrence of the fever, this is clearly the ontic sense of explanation, and the word "cause" can be substituted for "explain" without change of meaning.

In bare causal explanations, the underlying explanatory relation is causal dependence. But to show that this dependence exists is to offer a very partial model of this causal relationship. Figure 8.1 shows what the model looks like.

This is not much of a model—it is the sketchiest of mechanism sketches—but nonetheless it is a representation that shows a pattern of dependence. If, in a particular case, this model bears a sufficient resemblance relation to the target (e.g., the process by which Clara contracted puerperal fever), then this is a minimally explanatory model.

Bare causal explanations identify causal dependencies without describing the mechanisms that are responsible for those dependencies. But bare causal explanations need not be as simple as this. Epidemiological data about risk factors for diseases will, for instance, describe many factors that may be causally relevant to a disease. For diabetes these might include facts about age, weight, diet, exercise, blood pressure, etc. The epidemiological evidence is such that there is good reason to believe that all of these factors contribute in some way to the onset of diabetes. If these factors did contribute to the onset of diabetes in a particular case, then we would have a bare causal explanation with multiple causal factors. Such explanations remain bare causal explanations so long as they provide no account of the mechanisms and processes by which these factors contribute to the production of the effect.

While I treat bare causal explanations as a discrete category, there is no bright line between bare causal explanations and mechanistic explanations. The key distinction between the two is that bare causal explanations simply assert the existence of causal dependencies, arising from unspecified mechanisms, while mechanistic explanations describe the mechanisms that are responsible for those dependencies. But, as Darden, Craver, and others have emphasized, rough or sketchy mechanistic models are gradually filled in (Machamer, Darden, and Craver 2000; Glennan 2005; Craver and Darden 2013). As one identifies more things upon which a phenomenon depends,

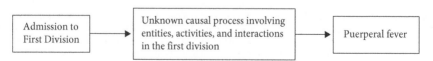

Figure 8.1 A simple model of causal dependencies.

and as one begins to describe how these things are organized so as to collectively bring about the phenomenon, one begins to get a model of a mechanism.[16]

The explanation of Clara's puerperal fever is an example of a bare causal explanation that is singular. But such explanations are also offered for more general kinds of causal relationships. Let us consider how this works. When Semmelweis sought an explanation of the causes of the epidemic at the Vienna General Hospital, one might wonder whether he was seeking a singular or a general explanation. At first glance, it appears to be a general explanation, because he was not seeking to explain any particular case of childbed fever. But really, his primary goal was to explain the high rate of puerperal fever in a certain population—namely, expecting mothers admitted to the hospital, and specifically to its First Division. And a population like the one in the First Division really is an individual. It is spatially and temporally localized, and the women who make it up are not instances of some abstract class, but rather parts of a complex individual.[17] If Semmelweis's claim that contact with dirty-handed interns explained the increased incidence of fever in the First Division is true, it is because a large percentage of the cases of childbed fever actually were produced by an infection introduced through an examination by a dirty-handed intern.[18]

While Semmelweis's immediate concern was with the situation at the Vienna General Hospital in the 1840s, his conclusions are of much wider application. Put rather generally, the conclusion that can be drawn from Semmelweis's work is that contact with unsterilized body parts or other objects like medical instruments can cause sepsis. But does this vague claim count as a general bare causal explanation?

I would argue that until this generalization is attached to particular cases, there is no explanation. Causation requires mechanisms, and mechanisms are spatiotemporally localized particulars. There are no infections or body parts in general, only particular infections or particular body parts. What we have with the generalization by itself is only an unattached model—one that could be deployed to explain many particular targets. What Semmelweis discovered was that a certain kind of mechanism could produce a certain kind of disease. Infections can be transferred from one person to another by contact, and these infections can produce the set of symptoms we associate

[16] Woodward (2011) has suggested that mechanistic explanation just involves fine-grained difference-making information. While I disagree with Woodward that causation can be reduced to difference-making, and I would perhaps emphasize more than he how non-causal forms of organization matter, I think he is fundamentally correct that basic causal explanations identify dependencies, and that mechanistic explanations fill in.

[17] See Millstein (2009) for a discussion of reasons to treat populations as individuals. While her account is motivated primarily by considerations in evolutionary biology, similar considerations apply here.

[18] The exact truth conditions are vague here, because the explanatory claim is vague and the model is not completely specified. It is important to remember that puerperal fever did not just occur in the First Division, but also in the Second Division, in other hospitals, and in locations outside of hospitals. Thus, it need not be the case that all cases in the First Division were caused by these examinations. Ideally one might expect the examinations to account for all of the cases beyond some base rate, but so long as a substantial portion of the differential was due to this, one would I think count Semmelweis's explanation as a legitimate one.

with sepsis. But if the account of mechanism kinds (and kinds more generally) is only weakly realist, there really is no explaining of kinds per se. There is rather only a causal schema of wide applicability that can be deployed into contexts where it can do explanatory work.

8.5   Mechanistic Explanations

Mechanistic explanations are how explanations. They explain how some phenomenon comes about via a model of the mechanism that is responsible for it. In Chapter 3, I offered an account of what mechanistic models are and how they explain. Here I want to focus on how mechanistic explanations are related to bare causal and non-causal explanations.

A model is mechanistic when elements of the model can be mapped onto the mechanism whose behavior the modeler seeks to explain. David Kaplan puts this condition in terms of what he calls the mechanism-model-mapping (3M) constraint:

(3M) A model of a target phenomenon explains that phenomenon to the extent that (a) the variables in the model correspond to identifiable components, activities, and organizational features of the target mechanism that produces, maintains, or underlies the phenomenon, and (b) the (perhaps mathematical) dependencies posited among these (perhaps mathematical) variables in the model correspond to causal relations among the components of the target mechanism.   (Kaplan 2011)

Kaplan offers his 3M constraint in an effort to delineate descriptive from explanatory models in computational neuroscience, and as such, he offers 3M as a general constraint on explanation. Put this way, 3M implies that the only explanations are mechanistic explanations. I do not want to construe 3M in this way, as I am arguing that there are other ways to explain besides mechanistically. So for my purposes it is best to amend Kaplan's constraint by introducing the word "mechanistically" into a modified constraint (3M*), which states that "a model of a target phenomenon *mechanistically* explains that phenomenon to the extent that (a)...and (b)..."

One reason it is important to distinguish between mechanistic explanation and explanation as such is that a phenomenal (or descriptive) model may contain information about bare causal dependencies, and hence may explain. Consider the phenomenon of vowel recognition. Speech sounds, including vowel sounds, are principally characterized by a feature of the acoustic signal called formants. Formants are local maxima in the intensity of speech signals at given frequencies in the acoustic spectrum, and speakers can alter the frequencies of these formants by changing the position of parts of their vocal tract. It turns out that listeners recognize vowel sounds chiefly by the frequency of the first two formants, which form a two-dimensional "vowel space." A visual representation of this space is given in Figure 8.2.[19]

---

[19]   This figure is based on data from Gerstman (1968). Glennan (2005) contains a case study of mechanistic models of vowel normalization from which this example is drawn.

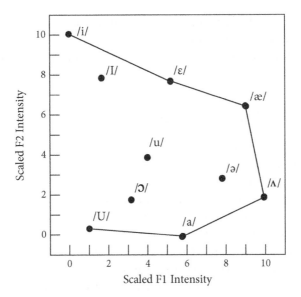

Figure 8.2 An average talker's vowel space.

The mapping in Figure 8.2 is a description of the behavior of human vowel recognition mechanisms. It does not meet the constraints of 3M*, since it contains no information on what the components of the responsible mechanism are or how they interact to produce the phenomenon described. Nonetheless, the model contains explanatory information. It shows what features of the speech signal vowel recognition depends upon. It allows one to answer questions about what would happen if the speech signal changed—Woodward's (2003) what-if-things-had-been-different questions. Thus, this mapping allows one to offer a bare causal explanation of, e.g., why a speaker hears a certain signal as a particular vowel. A mechanistic explanation answers a different explanatory question. In this case, a mechanistic explanation describes how the signal is processed so as to detect the formants. This involves both identifying the activities involved in speech processing and trying to locate the entities within the auditory system that are responsible for these activities.

Kaplan's 3M requirement can been read as suggesting completeness is an explanatory ideal—what I have called the ideal-model model and Chirimuuta (2014) has called the more details the better (MDB) assumption. As I have already indicated in Chapter 3, I do not agree with MDB. Abstraction and idealization are not just epistemic concessions; they allow us to separate principal from secondary causes within mechanisms and they allow us to achieve generality in our representations. The degree to which more is better depends upon the kind of explanatory target and explanatory aims.

The key point for our discussion of explanation is that abstract, idealized, and incomplete models can satisfy the 3M constraint, and thus can be used to explain how.

This implies that the distinction between mechanistic and non-causal explanations cannot be cashed out in terms of distinctions between different levels of abstraction or completeness. Non-causal explanations are typically abstract, but that is not what make them non-causal.

8.6 Non-Causal Explanation

A central claim of this chapter is that the fact that natural phenomena have causes does not imply that all explanations of these phenomena are causal. This is because explanation requires dependence, and there are non-causal forms of dependence, even in the natural world. To substantiate this claim we need to give a more detailed and positive characterization of what non-causal explanations are. This is made difficult by the dizzying variety of explanations that philosophers purport to be non-causal.[20] As I sketch a general account of non-causal explanation, I do not want to suggest that all non-causal explanation has one form. What I will try to do is offer a metaphor that can help draw a line between non-causal explanations and the causal and mechanistic ones.

The key idea is that all causal mechanisms act within some space, and that sometimes the best way to explain how these mechanisms behave or where they end up is not to trace the path of mechanisms through the space but rather to show how the structure of that space (its geometry or topology) constrains or shapes the behavior of those mechanisms.[21] Space here is a metaphor that can be cashed out in various ways to yield different varieties of non-causal explanation. The space might be physical space-time itself, with its distinctive geometry; or it might be a social space, or a fitness landscape, or indeed the state space of any dynamical system.

Here is a simple non-causal explanation that illustrates this idea. Imagine a uniformly curved bowl, and a marble held at any point on the bowl's interior surface. If the marble is released, it will begin to roll back and forth from one side of the bowl to another. If the bowl were a frictionless surface, the marble would oscillate forever back and forth between two points on opposite sides of the bowl; but, because of friction, the marble will slowly lose energy, traveling each time a bit less far up the side of the bowl until eventually it comes to rest at the bottom of the bowl.

How shall we explain where the marble ends up? One way would be to calculate the trajectory of the marble as its velocity and acceleration change due to the operations of gravitational, frictional, and normal forces. One could follow this trajectory from the

---

[20] A partial list includes equilibrium explanations (Sober 1983), structural explanations in physics (Dorato and Felline 2011; Hughes 1989), distinctively mathematical explanations (Lange 2013), minimal explanations (Batterman 2001; Batterman and Rice 2014), design explanations (Wouters 2007), topological explanations (Huneman 2010; 2015), effective coding explanations (Chirimuuta 2014), and (on certain interpretations) evolutionary explanations that appeal to fitness (Matthen and Ariew 2002).

[21] Huneman's (2010) account of topological explanations and their relation to mechanistic explanations appeals to the metaphor of space, and my thinking owes much to his remarks.

release of the ball until it finally comes to rest. This would be a mechanistic explanation that shows exactly how the marble got to where it ended up.

On the other hand, it is also possible to explain the final resting place of the marble by reference only to the shape of the bowl and some basic facts about the setup (like the fact that the bowl and marble are smooth enough to allow the marble to roll and that the bowl is resting flat upon the earth, i.e., so the plane of the bottom of the bowl is perpendicular to the direction of the gravitational field). No matter where the marble starts, it will always end up at the bottom. We do not need to know exactly where it started or exactly how it traveled there.

The marble in the bowl is a simple example of what Elliot Sober (1983) calls an equilibrium explanation. Equilibrium explanations in general show how dynamical systems with a specified causal organization will, within some range of parameter values, always reach some equilibrium state. Equilibrium explanations show what the equilibrium depends upon—in this case, the setup with the bowl, the marble, and the gravitational field—and also what it does not depend upon—in this case, the point along the bowl's surface at which the marble is released.

One might argue here that equilibrium explanations are actually causal explanations, because features like the shape of the bowl and the existence of the gravitational field are in fact causally relevant to the trajectory and endpoint of the marble. And indeed there is a causal explanation in the neighborhood that would appeal to just these factors. But what the equilibrium explanation does that makes it non-causal is that it describes the features of the space that constrain the causal process, rather than describes the causal process itself.

Equilibrium explanations may be preferred over causal and mechanistic explanations for a number of reasons. For one thing, they are economical, since they allow exclusion of a lot of information about the system that is relevant to its trajectory but not to its endpoint. One can in principle explain why a certain marble ends up at the bottom of the bowl by constructing a model that traces its trajectory from a certain point—but such details do not make a difference to where the marble ends up.

Equilibrium explanations also unify. Because an equilibrium explanation provides a model that specifies ranges for one or more parameters that are irrelevant to the dynamical system's equilibrium state, any actual system that can be characterized by that model and whose parameters fall within those ranges can be explained by that model. The opportunities for unification in this instance go beyond simply unifying a collection of different releases of a marble from a bowl. The same basic model can be applied to bowls of a wide variety of shapes (roughly so long as they are concave), and more generally, the system is an example of a damped oscillator, of which there are many kinds that involve neither marbles nor bowls.

Sober's primary example of an equilibrium explanation was R. A. Fisher's explanation of the 1:1 sex ratio. Sober summarizes Fisher's explanation as follows:

If a population ever departs from equal numbers of males and females, there will be a repro-ductive advantage favoring parental pairs that over-produce the minority sex. A 1:1 ratio will be the resulting equilibrium point. The ratio of male to female progeny has an impact on a parent's fitness in virtue of the number of grandchildren that are produced. If males are now in the majority, an individual who produces all female offspring will on average have more grandchildren than one that produces all males or a mixture of sons and daughters.

(Sober 1983, 201–2)

Like the marble in the bowl, this explanation provides an account of where something will end up that is independent of the actual path it takes. Here the space is not a physical space and the something is not a physical object like a marble. Instead, the something is a feature of a population, the sex ratio. The sex ratio may vary, but over time, natural selection will, like the forces of gravity and friction, draw it toward a stable equilibrium point.

Like the marble in the bowl explanation, Fisher's explanatory model can be applied to a wide range of cases. The explanation's power derives in part from how much detail can be left out. Fisher's explanation should hold for a wide range of sexually reproducing species, and is independent of the details of the reproductive cycle, procedures for mate selection, and the genetic (or epigenetic, or cultural) basis of the sex determination mechanism.

While equilibrium explanations like that of the marble in the bowl or the explanation of the sex ratio are not mechanistic, they are nonetheless explanations of mechanism-dependent phenomena. Consequently, they can only explain if assumptions about those mechanisms that are implicit or explicit in the explanatory model do in fact hold true of the target mechanisms. Consider again Fisher's explanation of the sex ratio. While the account can explain the phenomenon independently of most details of the processes of reproduction, many assumptions must in fact hold for the result to obtain. A fundamental assumption is that the causes of any bias in the ratio of offspring will have a basis that is heritable, and hence subject to selection. The result also assumes that the parental investment to produce and sustain males and females is equal. If, say, parents could produce two females for the price of one male, then you would not get a 1:1 sex ratio. A third assumption is that mating is random across the global population, and the list does not end here.[22]

Let us consider another kind of non-causal explanation in evolutionary biology, what Arno Wouters (2007) has called *design explanation*. Wouters claims that an important form of explanatory reasoning in biology involves explanations of why organisms of a certain kind have one trait rather than another. For instance, why do fish have gills rather than lungs (and why conversely do terrestrial tetrapods, including amphibians, reptiles, mammals, and birds, have lungs rather than gills); or why do both terrestrial tetrapods and fishes have specialized organs (lungs or gills) for respiration, rather than just breathing through the skin.

---

[22] Various assumptions are discussed in the philosophical literature in, among other places, Sober (1983) and Batterman and Rice (2014), which in turn rely on standard discussions from biology, especially from the work of Hamilton—which places Fisher's account within the context of evolutionary game theory. In treating this as an equilibrium explanation, I am relying on Sober's classic analysis.

According to Wouters's analysis, design explanation is a form of non-causal explanation, which reveals "constraints on being alive." These constraints are design constraints imposed by the organism's environment. The constraints that explain why fish have gills rather than lungs, while mammals and other terrestrial tetrapods have lungs rather than gills, have to do with basic facts about the different characteristics of water and air, and how these characteristics affect the rate of diffusion of oxygen across membranes. The basic physical principle involved is Fick's law of diffusion, which shows how the rate of diffusion across a membrane depends upon three parameters— a diffusion gradient representing the permeability of the membrane, the surface area of the membrane, and the concentration gradient, which depends upon how much of the diffusing gas (here oxygen) lies on each side of the membrane. Wouters goes on to explain in some detail how, given this and other physical principles, gills can only work in water while lungs can only work in air:

[B]iologists compare the physical qualities of water and air. The concentration of oxygen in air is about 30 times higher than in air-saturated water, the diffusion coefficient of oxygen is about 30,000 times higher in air than in water, water is more than 800 times as dense as air, and water is about 60 times more viscous than air. Animals that obtain oxygen from water have to solve three problems. First, due to water's low oxygen content, they need to ventilate a much larger volume than air breathers to extract the same amount of oxygen. Second, uptake of oxygen from water is very inefficient, due to both its lower oxygen content...and lower diffusion coefficient...Third, because of water's higher density and viscosity, it costs much more energy to move a certain volume of water over a certain distance than to move the same volume of air over that distance. Respiration in air faces other difficulties, including the constant risk of desiccation and the problems caused by gravitation. These problems do not occur in water.

The differences in design between gills and lungs are explained by these physical differences and the different problems that "result" from them. The respiratory surface of gills...is immensely larger than that of lungs, and gills' respiratory membranes are much thinner.... This increases oxygen uptake efficiency. The unidirectional character of the water's flow across the gills saves the costs of reversing the flow. In addition, the continuous stream of water over the respiratory surface increases the oxygen uptake because it increases the oxygen concentration at the outside of the respiratory surface. Due to the problems of gravitation, the vast increase in gills' surface area is not possible on land; the finely divided and thin lamellae would collapse against each other. The thin membranes and unidirectional flow also would increase the risk of desiccation.   (Wouters 2007, 70–1)

This explanation obviously cites all sorts of causally relevant features of oxygen, water, air, gills, and lungs, but it is not causal explanation. The reason, Wouters thinks, is that the explanation does not describe the causal process by which lungs and gills evolved, but rather identifies constraints on the possible products of those processes. We can explain the non-causal character of the explanation by appealing again to the metaphor of space. The space here is not a physical space but a design space. Lungs and gills in their many forms are solutions to a design problem—in this case, the problem of adding oxygen to and removing carbon dioxide from the bloodstream. A design explanation, on Wouters's account, shows how the environment places certain constraints on where in that design space an organism within that environment will end up. It must end up somewhere in the design space where the designs meet essential requirements for the organism staying alive.

A design explanation is similar to an equilibrium explanation in that it explains the outcome of a process without tracing the process. The process in the case of design explanation is an evolutionary process—the process by which the solution to the design problem evolved. We can usefully contrast this form of explanation with a mechanistic form of evolutionary explanation, which Calcott has called lineage explanations. As he puts it in his abstract:

> The aim of these explanations is to make plausible certain trajectories of change through phenotypic space. They do this by laying out a series of stages, where each stage shows how some mechanism worked, and the differences between each adjacent stage demonstrates how one mechanism, through minor modifications, could be changed into another. These explanations are important, for though it is widely accepted that there is an "incremental constraint" on evolutionary change, in an important class of cases it is difficult to see how to satisfy this constraint.   (Calcott 2009, 51)

Whereas in design explanations, the path through the design (i.e. phenotypic) space is ignored, in lineage explanations the point just is to trace that path. Instead of explaining why we got here in terms of constraints on where we had to end up, we explain why we got here by showing how we got here, i.e., by tracing the path.[23]

While the examples I have discussed so far all involve structural features of a space that direct causal processes toward some endpoint or equilibrium, not all kinds of non-causal explanations have this form. Consider the much-discussed example of the bridges of Königsberg. Königsberg (now Kalingrad, Russia) was, in the eighteenth century, a Prussian city that spanned the Pregel River. The city spread across both banks and two islands in the center of the river, and was connected by seven bridges. The mathematical problem raised by the bridges of Königsberg was whether it was possible, starting and ending at any point, to traverse all seven bridges exactly once. The mathematician Leonhard Euler showed that it was not. Euler's explanation begins with the construction of a model of the city and its bridges. The crucial topological properties of the city and its bridges can be represented using a graph like that in Figure 8.3. Here the nodes represent landmasses and the edges represent bridges.

---

[23] Let me briefly mention one other sort of explanation—functional explanation. On so-called selected effects theories of functions (Millikan 1984; Neander 1991), the presence of a trait in an organism is explained by the role that the trait played in enhancing the fitness of ancestral organisms. So, for instance, the reason why fish have gills is that ancestral gills had effects that were favored by natural selection. This looks at first glance to be a causal explanation, as the present existence of the trait depends upon the selective advantage to ancestors in the lineage who bore past versions of this trait. The explanation is certainly not a mechanistic one, as it does not show (like a lineage explanation) how a lineage evolved so as to have that trait. Arguably these explanations share with equilibrium explanations and design explanations an emphasis on the space through which evolutionary processes move, rather than the particular trajectory they follow. On the relation between design and selected effects explanations, see Wouters (2007).

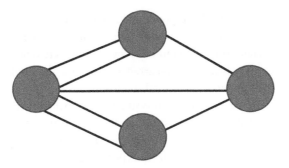

Figure 8.3 Graph of the bridges of Königsberg.

Given the conditions specified in the problem (e.g., no boats, no swimming, and no crossing the bridge halfway), these are the only features upon which Euler's answer depends.

Euler's crucial observation was that, other than the start and endpoints, the number of times you enter a node (i.e., one of the landmasses) must equal the number of times you leave it. This means that if you must cross each edge (bridge) exactly once, the number of edges for any nodes in the middle of the walk must be even. If the start node has an even number of edges, you must end your walk where you start it. If the start node has an odd number of edges, there must be exactly one other node with an odd number of edges, where your walk will end. The number of edges connected to a node is called its degree, and the generalized theorem is that a walk that traverses each edge (an Euler walk) is only possible for graphs where either zero or two nodes have odd degree. In the original problem, there are four nodes, all of odd degree, which is why no Euler walk is possible.

The explanation of why the bridges of Königsberg could not be traversed exactly once shares a number of features with equilibrium explanations. Like equilibrium explanations, Euler's explanation shows how structural features of a space entail a result that is independent of any particular causal process in that space. In addition, like these other explanations, it is highly generalizable. Euler's result can be applied to any finite network that can be characterized in terms of nodes and edges, not just to Königsberg, or to bridges or walks. Nonetheless, it is not an equilibrium explanation because it does not show where actual or possible causal processes end up. It offers a different sort of constraint on the possible: that no actual or possible causal processes in certain graphs can follow an Eulerian path.

Among causal explanations we drew a distinction between bare causal explanations, which show what depends upon what, and mechanistic explanations, which show why and how those dependencies obtain. We can draw a similar distinction between varieties of non-causal explanation. To illustrate, consider a variation on the bridges of Königsberg. Suppose that the town of Xanadu was situated, like Königsberg, on the islands and banks of a river, and the citizens were similarly vexed by their inability to cross all bridges exactly once. A citizen of Königsberg, visiting Xanadu, is asked if he understands why this is so, and he responds that Xanadu has more than two landmasses with an odd number of bridges to them. He cannot remember the argument why it is so, but a mathematician from his country has shown that there is no way to cross each bridge exactly once if the bridges are so arranged. In citing the result without having the proof, has the visitor thereby explained anything?

I would argue yes, but not much. The situation is analogous to the case where Semmelweis had evidence that unwashed hands were a cause of childbed fever but where he had no account of why that connection obtained. In that case, I argued, he hypothesized correctly that the rise of fever depended upon the lack of hand-washing and so at least began to explain why women in the Second Division had a higher rate of fever. A bare causal explanation is explanatory when it correctly identifies a dependence relation, even when it does not identify the causal process in virtue of which that dependence relation obtains. In the "mere non-causal" explanation, a certain fact could be explained by a structural dependency (here the dependence of non-Eulerian walks on too many nodes of odd degree), even if why this dependency exists has not been explained.

My examples so far have characterized non-causal explanations as explanations in which the structure of a space constrains the activities of mechanisms operating in that space. These spaces can be physical spaces, as in the city of Königsberg, design spaces as discussed by Wouters, or, more generally state spaces of systems. There are, however, some non-causal explanations in which the explanations are not explanations about a space in which particular kinds of mechanisms operate, but rather explanations that fall out of the basic physical structure of the universe, the space in which all processes of any kind operate. I have in mind here what are sometimes called structural explanations in physics (Dorato and Felline 2011; Hughes 1989). Consider, for instance, the related phenomena of length contraction and time dilation. Suppose that there are two identical spaceships (with identical lengths and identical clocks) that are at rest with respect to each other. If the ships take off so that they are in motion relative to each other, observers on each ship will perceive the clocks on the other ship to slow down, and the length of the other ship to contract. The explanations of these phenomena are geometrical. They fall out of features of the space-time distance metric in Minkowski space-time.[24] These phenomena are in no way causal or mechanistic. Although some sort of mechanism would be required to accelerate one of the spaceships so that it was in motion relative to the other, this acceleration does not cause time to slow or compress the rocket. Observers on rockets in relative motion will observe these phenomena, even though nothing is happening. Measurements of the activities of any mechanisms whatsoever will be subject to these relativistic "effects"—though they are not really effects, since they are not caused, and they are at any rate negligible except at relative speeds approaching the speed of light.

---

[24] I shall not rehearse here the geometrical explanation of these phenomena. See Felline (2015) for a succinct review that highlights the non-causal and non-mechanistic character of length contraction.

8.7   Conclusion: What Makes Explanations Causal or Non-Causal

In light of the examples of explanations discussed in this chapter, let us take stock of the general conclusions we can draw about the nature of explanation, and of how to draw the line between causal and non-causal varieties. To begin, I have argued that that the sine qua non of explanation is dependence. Explanations are models (including equations, simulations, physical models, diagrams, or other forms of representation) that exhibit dependencies that exist in the parts of the natural world that those models are used to represent. Explanatory models may unify, and that is an epistemic virtue, but not one required for explanation.

Given that dependence is the sine qua non of explanation, my aim in describing various examples of causal and non-causal explanation has been to characterize the different kinds of dependence relations involved in each. The key to sorting these different varieties of dependence out is to recognize that all mechanisms and causal processes occur within some space. Causal processes or mechanisms operate within spaces of many kinds—space-time itself, physical spaces like bowls or race-car tracks, social spaces, phenotypic spaces, or state spaces of all descriptions. Those spaces will inevitably have a certain structure that will shape the directions in which the causal processes go. If we were to look for a generic label that applies to all non-causal explanations, we might call them structural explanations. Although the term "structural explanation" has been used by philosophers as well as natural and social scientists in ways that are not wholly consistent, it is a suggestive term because it reminds us that the spaces in which causal mechanisms operate have structure, and this structure can sometimes better explain why things are as they are than explanations that appeal to these mechanisms.

This is not the only way we might try to draw the distinction between the causal and the non-causal. Sometimes, for instance, it is suggested that non-causal explanations are more abstract than causal explanations. Perhaps this tends to be so, but this is not a good way to draw a distinction, because many models that are causal (and mechanistic) can be both abstract and idealized. Also, one might read certain authors as suggesting that some explanations are non-causal because they appeal to mathematical rather than causal dependencies. But one cannot really infer anything about whether a model is causal from the mere fact that it uses mathematics. Any mathematical model will include mathematical dependencies between parameters or variables, and some of these mathematical dependencies will represent causal dependencies. A mathematical model of an oscillating spring, for instance, will have a mathematical expression describing the linear restoring force as the spring moves from its rest position. The linear mathematical dependency will represent a causal dependency. At the same time, it seems possible to model non-causal dependencies with models that do not use mathematics at all. For instance, a diagram or physical model of a system might exhibit dependencies that are not mathematically characterized but which are not causal. The diagram of the bridges of Königsberg provides an example. The bottom line is that mathematical dependencies are dependencies in our models rather than in the world. And whether a mathematical dependency in a model represents a causal or non-causal form of explanatory dependence is determined not by the model itself, but by what the mathematical dependency is taken to represent.

And here is a final point about mechanistic explanation. I have classified mechanistic explanations, following the usual custom, as causal, and so to a first approximation they are. But mechanistic explanations exhibit both causal and constitutive forms of dependence. Mechanisms are responsible for their phenomena because of the organized activities and interactions of their parts. Organization, even causal organization, depends upon non-causal relationships—where things are in space and time; how they are structured, and how wholes are made of parts. So it is not just the case that explanations of mechanisms are complemented by non-causal explanations of the space in which mechanisms act. Mechanisms themselves depend upon non-causal organization as well as causal activities and interactions.

Postscript

As I have worked on this book, I have often been struck by the audacity of its title. The phrase "the New Mechanical Philosophy" was not my own, but still, to appropriate the venerable tradition of mechanical philosophy, to call it new, and to add a definite article to boot—maybe that's just too much. Nonetheless I have stuck with the title, and the project, because I think it reflects both the continuity with the history of science and its philosophy, and a sea change in philosophical thinking in the new century. Mechanical philosophy is as old as Democritus; it was a central theme in the scientific revolution; it has helped drive research and debates on the nature of life and the nature of mind. But the New Mechanical Philosophy is new in large part because it tracks changes in the way science is done. Over recent decades, the sciences have developed increasing, if still rudimentary, capacities to analyze complex and heterogeneous systems—cells, brains, ecosystems, economies, and so on. Whereas in earlier epochs many of the greatest scientific achievements have been to understand the basic building blocks—the laws of electricity and magnetism or the structure of the hydrogen atom—scientists are now able to greater and greater extents understand how these things are put together to make the universe we know. Mechanical philosophy is always about understanding how things are put together, so it is a philosophy for this time.

By now, the New Mechanical Philosophy is not so new. It has its proponents and critics, and its stock of debates—many of which I have tried to engage with in this book. I would like, though, to conclude with a few words about what I hope is next for those interested in continuing the conversation.

One place I hope the discussion will go is toward a more serious and continual engagement with traditional questions of metaphysics. Born of its concern with scientific practice, and imbued with a touch of empiricist disdain for the traditional metaphysical enterprise, the New Mechanism has often avoided or glossed over issues of ontology and metaphysics. I hope, however, I have shown that further progress in understanding the nature and consequences of the mechanistic approach to science will depend upon our thinking carefully about its ontological implications. This will involve further exploration of questions traditionally within the purview of metaphysics—the nature of substances, objects, properties, causes, laws, powers, dispositions, and modality. There is also considerable opportunity to explore further some traditional metaphysical debates within the philosophy of science—over emergence or scientific realism, for instance—from the perspective of the New Mechanism. And, while I know that philosophers of science like me can learn from our colleagues in metaphysics, I hope too that the New Mechanism can offer metaphysicians both new problems and new ways of thinking about old ones. Above all, I hope that thinking about mechanisms metaphysically may spur a change in direction away from questions of what basic things there are in the world to compositional questions about how things hang together.

I hope too that the strategy embodied in minimal mechanism will increasingly take root, and that the discussion will move from questions about what is and is not a mechanism to a more deliberate exploration of the variety of mechanisms. Philosophers, like scientists, are both empowered and limited by their models. Mendel's peas and Morgan's fruit flies were responsible for the birth of genetics, but continuing developments in that field have shown that those model organisms provided only the most rudimentary sense of the great variety of mechanisms of inheritance. So, too, it is reasonable to think that the stock of examples that have driven recent philosophical thinking about mechanisms—cellular respiration, protein synthesis, the action potential, or, for that matter, toilets—have both elucidated our understanding of mechanisms, but also limited our vision both of the variety of mechanisms and of the diversity of methods needed to discover, represent, and explain them. While the exploration of the varieties of mechanisms will certainly help us understand methodological differences across scientific fields, the commonalities we find will help us understand how methods may be shared and disciplines be integrated. In short, exploring the variety of mechanisms may help us understand both the unity and the diversity of science.

While I have framed these remarks as suggestions for future research, a review of literature from the last five years—much of it cited in this book—suggests that both a more serious metaphysical engagement and an exploration of mechanistic variety are well underway.[1] Within the coming years I fully expect the term "New Mechanism" to fade from view, but I think the mechanistic turn in both science and its philosophy is here to stay. That is just how we have found the world to be.

---

[1] The issues are also central in the chapters of a forthcoming anthology on mechanisms and mechanical philosophy that I have edited with Phyllis Illari (Glennan and Illari 2018).

References

Abrams, Marshall. 2012. "Mechanistic Probability." *Synthese* 187 (2): 343–75. doi:10.1007/s11229-010-9830-3.

Abrams, Marshall. 2018. "Probability and Chance in Mechanisms." In *The Routledge Handbook of Mechanisms and Mechanical Philosophy*, edited by Stuart Glennan and Phyllis Illari, 169–84. London: Routledge.

Aizawa, Kenneth, and Carl Gillett. 2009. "The (Multiple) Realization of Psychological and Other Properties in the Sciences." *Mind & Language* 24 (2): 181–208.

Allen, Gar. 2018. "Mechanism, Organicism and Vitalism." In *The Routledge Handbook of Mechanisms and Mechanical Philosophy*, edited by Stuart Glennan and Phyllis Illari, 59–73. London: Routledge.

Andersen, Holly. 2011. "Mechanisms, Laws, and Regularities." *Philosophy of Science* 78 (2): 325–31.

Andersen, Holly. 2012. "The Case for Regularity in Mechanistic Causal Explanation." *Synthese* 189 (3): 415–32. doi:10.1007/s11229-011-9965-x.

Andersen, Holly. 2014a. "A Field Guide to Mechanisms: Part I." *Philosophy Compass* 9 (4): 274–83.

Andersen, Holly. 2014b. "A Field Guide to Mechanisms: Part II." *Philosophy Compass* 9 (4): 284–93.

Anscombe, G. E. M. 1993. "Causality and Determination." In *Causation*, edited by Ernest Sosa and Michael Tooley, 88–104. Oxford Readings in Philosophy. Oxford: Oxford University Press.

Ariew, André. 2008. "Population Thinking." In *Oxford Handbook of Philosophy of Biology*, edited by Michael Ruse, 64–86. Oxford: Oxford University Press.

Armitage, P., and R. Doll. 1954. "The Age Distribution of Cancer and a Multi-Stage Theory of Carcinogenesis." *British Journal of Cancer* 8 (1): 1–12.

Armstrong, David M. 1983. *What Is a Law of Nature?* New York: Cambridge University Press.

Ayala, Francisco J., and Cathryn A. Campbell. 1974. "Frequency-Dependent Selection." *Annual Review of Ecology and Systematics* 5: 115–38.

Baetu, Tudor M. 2013. "Chance, Reproducibility, and Mechanistic Regularity." *International Studies in the Philosophy of Science* 27 (3): 253–71. doi:10.1080/02698595.2013.825492.

Baetu, Tudor M. 2018. "Mechanisms in Molecular Biology." In *The Routledge Handbook of Mechanisms and Mechanical Philosophy*, edited by Stuart Glennan and Phyllis Illari, 308–18. London: Routledge.

Bajec, Iztok Lebar, and Frank H. Heppner. 2009. "Organized Flight in Birds." *Animal Behaviour* 78 (4): 777–89. doi:10.1016/j.anbehav.2009.07.007.

Batterman, Robert W. 2001. *The Devil in the Details: Asymptotic Reasoning in Explanation, Reduction, and Emergence*. New York: Oxford University Press. doi:10.1093/0195146476.001.0001.

Batterman, Robert W., and Collin C. Rice. 2014. "Minimal Model Explanations." *Philosophy of Science* 81 (3): 349–76. doi:10.1086/676677.

Baumgartner, Michael, and Alexander Gebharter. 2016. "Constitutive Relevance, Mutual Manipulability, and Fat-Handedness." *The British Journal for the Philosophy of Science* 67 (3): 731–56.

Bautista Paz, Emilio, Marco Ceccarelli, Javier Echávarri Otero, and Jose Luis Muñoz Sanz. 2010. *A Brief Illustrated History of Machines and Mechanisms.* Dordrecht: Springer Netherlands.

Beatty, John. 1995. "The Evolutionary Contingency Thesis." In *Concepts, Theories, and Rationality in the Biological Sciences,* edited by Gereon Wolters and James G. Lennox, 45–82. Pittsburgh: University of Pittsburgh Press.

Bechtel, William. 2008. *Mental Mechanisms: Philosophical Perspectives on Cognitive Neuroscience.* New York: Routledge.

Bechtel, William. 2009. "Looking Down, Around, and Up: Mechanistic Explanation in Psychology." *Philosophical Psychology* 22 (5): 543–64.

Bechtel, William. 2010. "The Downs and Ups of Mechanistic Research: Circadian Rhythm Research as an Exemplar." *Erkenntnis* 73 (3): 313–28.

Bechtel, William. 2011. "Mechanism and Biological Explanation." *Philosophy of Science* 78 (4): 533–57.

Bechtel, William. 2012. "Understanding Endogenously Active Mechanisms: A Scientific and Philosophical Challenge." *European Journal for Philosophy of Science* 2 (2): 233–48. doi:10.1007/s13194-012-0046-x.

Bechtel, William, and Adele Abrahamsen. 2005. "Explanation: A Mechanist Alternative." *Studies in History and Philosophy of Science Part C: Studies in History and Philosophy of Biological and Biomedical Sciences* 36 (2): 421–41.

Bechtel, William, and Adele Abrahamsen. 2010. "Dynamic Mechanistic Explanation: Computational Modeling of Circadian Rhythms as an Exemplar for Cognitive Science." *Studies in History and Philosophy of Science* 41 (3): 321–33.

Bechtel, William, and Robert C. Richardson. 1993. *Discovering Complexity: Decomposition and Localization as Strategies in Scientific Research.* Princeton: Princeton University Press.

Bird, Alexander, and Emma Tobin. 2016. "Natural Kinds." In *Stanford Encyclopedia of Philosophy* (Spring 2016 edition), edited by Edward N. Zalta.

Boas, Marie. 1952. "The Establishment of the Mechanical Philosophy." *Osiris* 10: 412–541.

Bogen, Jim. 2005. "Regularities and Causality: Generalizations and Causal Explanations." *Studies in History and Philosophy of Science Part C: Studies in History and Philosophy of Biological and Biomedical Sciences* 36 (2): 397–420.

Bogen, Jim. 2008a. "The Hodgkin-Huxley Equations and the Concrete Model: Comments on Craver, Schaffner, and Weber." *Philosophy of Science* 75 (5): 1034–46. doi:10.1086/594544.

Bogen, Jim. 2008b. "The New Mechanical Philosophy." *Metascience* 17 (1): 33–41. doi:10.1007/s11016-007-9152-3.

Bogen, Jim. 2008c. "Causally Productive Activities." *Studies in History and Philosophy of Science Part A* 39 (1): 112–23.

Bolker, Jessica. 1995. "Model Systems in Developmental Biology." *BioEssays* 17 (5): 451–5.

Bolker, Jessica. 2012. "Model Organisms: There's More to Life than Rats and Flies." *Nature* 491 (7422): 31–3.

Bontly, Thomas D. 2006. "What Is an Empirical Analysis of Causation?" *Synthese* 151 (2): 177–200. doi:10.1007/s11229-004-2470-8.

Booch, Grady. 1994. *Object-Oriented Analysis and Design with Applications*. 2nd ed. Redwood City, CA: Benjamin/Cummings Pub. Co.

Boyd, Richard. 1989. "What Realism Implies and What It Does Not." *Dialectica* 43 (1–2): 5–29.

Boyd, Richard. 1991. "Realism, Anti-Foundationalism and the Enthusiasm for Natural Kinds." *Philosophical Studies* 61 (1): 127–48.

Brigandt, Ingo, Sara Green, and Maureen A. O'Malley. 2018. "Systems Biology and Mechanistic Explanation." In *The Routledge Handbook of Mechanisms and Mechanical Philosophy*, edited by Stuart Glennan and Phyllis Illari, 362–74. London: Routledge.

Bunge, Mario. 1997. "Mechanism and Explanation." *Philosophy of the Social Sciences* 27 (4): 410–65.

Calcott, Brett. 2009. "Lineage Explanations: Explaining How Biological Mechanisms Change." *British Journal for the Philosophy of Science* 60: 51–78. doi:10.1093/bjps/axn047.

Calcott Brett. 2011. "Wimsatt and the Robustness Family: Review of Wimsatt's Re-Engineering Philosophy for Limited Beings." Biology and Philosophy 26: 281–93.

Callebaut, Werner. 1993. *Taking the Naturalistic Turn, or How Real Philosophy of Science Is Done*. Chicago: University of Chicago Press.

Callebaut, Werner. 2013. "Scholastic Temptations in the Philosophy of Biology." *Biological Theory* 8 (1): 1–6.

Campaner, Raffaella. 2013. "Mechanistic and Neo-Mechanistic Accounts of Causation: How Salmon Already Got (Much of) It Right." *Metateoria* 3 (February): 81–98.

Campaner, Raffaella, and Maria Carla Galavotti. 2007. "Plurality in Causality." In *Thinking about Causes: From Greek Philosophy to Modern Physics*, edited by Gereon Wolters and Peter Machamer, 200–21. Pittsburgh: University of Pittsburgh Press.

Cartwright, Nancy. 1983. *How the Laws of Physics Lie*. Oxford: Clarendon Press.

Cartwright, Nancy. 1989. *Nature's Capacities and Their Measurement*. New York: Clarendon Oxford Press.

Cartwright, Nancy. 1995. "'Ceteris Paribus' Laws and Socio-Economic Machines." *Monist* 78 (3): 276–94.

Cartwright, Nancy. 1999. *The Dappled World: A Study of the Boundaries of Science*. Cambridge, UK; New York: Cambridge University Press.

Casini, Lorenzo. 2016. "Can Interventions Rescue Glennan's Mechanistic Account of Causality?" *The British Journal for the Philosophy of Science* 67 (4): 1155–83.

Chalmers, David, David Manley, and Ryan Wasserman, eds. 2009. *Metametaphysics: New Essays on the Foundations of Ontology*. Oxford: Oxford University Press.

Chirimuuta, Mazviita. 2014. "Minimal Models and Canonical Neural Computations: The Distinctness of Computational Explanation in Neuroscience." *Synthese* 191 (2): 127–53. doi:10.1007/s11229-013-0369-y.

Coleman, James S. 1986. "Social Theory, Social Research, and a Theory of Action." *American Journal of Sociology* 91 (6): 1309. doi:10.1086/228423.

Colombo, Matteo, Stephan Hartmann, and Robert Van Iersel. 2015. "Models, Mechanisms and Coherence." *British Journal of Philosophy of Science* 66 (1) 181–212.

Couch, Mark B. 2011. "Mechanisms and Constitutive Relevance." *Synthese* 183 (3): 375–88. doi:10.1007/s11229-011-9882-z.

Craver, Carl F. 2004. "Dissociable Realization and Kind Splitting." *Philosophy of Science* 71 (5): 960–71. doi:10.1086/425945.

Craver, Carl F. 2006. "When Mechanistic Models Explain." *Synthese* 153 (3): 355–76. doi:10.1007/sll229-006-9097-x.

Craver, Carl F. 2007. *Explaining the Brain*. Oxford: Oxford University Press.

Craver, Carl F. 2008. "Physical Law and Mechanistic Explanation in the Hodgkin and Huxley Model of the Action Potential." *Philosophy of Science* 75 (5): 1022–33.

Craver, Carl F. 2009. "Mechanisms and Natural Kinds." *Philosophical Psychology* 22 (5): 575–94.

Craver, Carl F. 2013a. "The Ontic Account of Scientific Explanation." In *Explanation in the Special Sciences: The Case of Biology and History*, edited by Marie I. Kaiser, Oliver R. Scholz, Daniel Plenge, and Andreas Hüttemann, 27–52. Synthese Library. Springer Netherlands. doi:10.1007/978-94-007-7563-3_2.

Craver, Carl F. 2013b. "Functions and Mechanisms: A Perspectivalist View." In *Functions: Selection and Mechanisms*, edited by Philippe Huneman, 133–58. Dordrecht: Springer Netherlands. doi:10.1007/978-94-007-5304-4.

Craver, Carl F., and William Bechtel. 2007. "Top-Down Causation without Top-Down Causes." *Biology and Philosophy* 22 (4): 547–63. doi:10.1007/s10539-006-9028-8.

Craver, Carl F., and Lindley Darden. 2005. "Introduction (Special Issue-Mechanisms in Biology)." *Studies in History and Philosophy of Science Part C: Studies in History and Philosophy of Biological and Biomedical Sciences* 36 (2): 233–44. doi:10.1016/j.shpsc.2005.03.001.

Craver, Carl F., and Lindley Darden. 2013. *In Search of Mechanisms: Discoveries across the Life Sciences*. Chicago: University of Chicago Press.

Craver, Carl F., and Marie I. Kaiser. 2013. "Mechanisms and Laws: Clarifying the Debate." In *Mechanism and Causality in Biology and Economics*, edited by Hsiang-Ke Chao, Stu-Ting Chen, and Roberta L Millstein, 125–45. Dordrecht: Springer Netherlands. doi:10.1007/978-94-007-2454-9_7.

Craver, Carl F., and James G. Tabery. 2015. "Mechanisms in Science." In *Stanford Encyclopedia of Philosophy* (Winter 2015 edition), edited by Edward N. Zalta.

Cummins, Robert. 2000. "'How Does It Work?' vs. 'What Are the Laws?' Two Conceptions of Psychological Explanation." In *Explanation and Cognition*, edited by F. Keil and R. Wilson, 117–45. Cambridge, MA: MIT Press.

Damasio, Antonio R. 1994. *Descartes' Error: Emotion, Rationality and the Human Brain*. New York: G. P. Putnam's Sons.

Darden, Lindley. 2005. "Relations among Fields: Mendelian, Cytological and Molecular Mechanisms." *Studies in History and Philosophy of Science Part C: Studies in History and Philosophy of Biological and Biomedical Sciences* 36 (2): 349–71. doi:10.1016/j.shpsc.2005.03.007.

Darden, Lindley. 2008. "Thinking Again about Biological Mechanisms." *Philosophy of Science* 75 (5): 958–69.

Darden, Lindley, and Joseph Cain. 1989. "Selection Type Theories." *Philosophy of Science* 56 (1): 106–29.

Darden, Lindley, and Carl F. Craver. 2002. "Strategies in the Interfield Discovery of the Mechanism of Protein Synthesis." *Studies in History and Philosophy of Science Part C: Studies in History and Philosophy of Biological and Biomedical Sciences* 33 (1): 1–28.

Darden, Lindley, and Nancy Maull. 1977. "Interfield Theories." *Philosophy of Science* 44 (1): 43–64.

Davidson, Donald. 1967. "Causal Relations." *Journal of Philosophy*, 64 (21): 691–703. Reprinted in *Causation*, edited by Ernest Sosa and Michael Tooley. Oxford: Oxford University Press.

Davidson, Donald. 1969. "The Individuation of Events." In *Essays in Honor of Carl G. Hempel*, edited by Nicholas Rescher, 265–83. Dordrecht: Reidel.

Davidson, Donald. 1995. "Laws and Cause." *Dialectica* 49 (2–4): 263–80.

Davies, P. C.W. 2004. "Does Quantum Mechanics Play a Non-Trivial Role in Life?" *Bio Systems* 78 (1–3): 69–79. doi:10.1016/j.biosystems.2004.07.001.

DeAngelis, Donald L., and Wolf M. Mooij. 2005. "Individual-Based Modeling of Ecological and Evolutionary Processes." *Annual Review of Ecology, Evolution, and Systematics* 36 (1): 147–68. doi:10.1146/annurev.ecolsys.36.102003.152644.

Decock, Lieven, and Igor Douven. 2011. "Similarity after Goodman." *Review of Philosophy and Psychology* 2 (1): 61–75. doi:10.1007/s13164-010-0035-y.

Demeulenaere, Pierre. 2011. *Analytical Sociology and Social Mechanisms*. Cambridge, UK; New York: Cambridge University Press.

Dennett, Daniel C. 1991. "Real Patterns." *The Journal of Philosophy* 88 (1): 27–51.

DesAutels, Lane. 2011. "Against Regular and Irregular Characterizations of Mechanisms." *Philosophy of Science* 78 (5): 914–25.

DesAutels, Lane. 2016. "Natural Selection and Mechanistic Regularity." *Studies in History and Philosophy of Science Part C: Studies in History and Philosophy of Biological and Biomedical Sciences* 57: 13–23. doi:10.1016/j.shpsc.2016.01.004.

DesAutels, Lane. 2018. "Mechanisms in Evolutionary Biology." In *The Routledge Handbook of Mechanisms and Mechanical Philosophy*, edited by Stuart Glennan and Phyllis Illari, 296–307. London: Routledge.

DesChene, Dennis. 2001. *Spirits and Clocks: Machine and Organism in Descartes*. Ithaca, NY: Cornell University Press.

Dijksterhuis, E. J. 1961. *The Mechanization of the World Picture*. Oxford: Clarendon Press.

Dorato, Mauro, and Laura Felline. 2011. "Scientific Explanation and Scientific Structuralism." In *Scientific Structuralism*, Boston Studies in the Philosophy of Science, edited by Alisa Bokulich and Peter Bokulich, 161–76. Dordrecht: Springer Netherlands. doi:10.1007/978-90-481-9597-8.

Dowe, Phil. 1992. "Wesley Salmon's Process Theory of Causality and the Conserved Quantity Theory." *Philosophy* 59 (2): 195–216.

Dowe, Phil. 2000. *Physical Causation*. Cambridge: Cambridge University Press.

Dowe, Phil. 2001. "A Counterfactual Theory of Prevention and 'Causation' by Omission." *Australasian Journal of Philosophy* 79 (2): 216–26.

Dray, William H. 1957. *Laws and Explanation in History*. London: Oxford University Press.

Ducasse, Curt J. 1993. "On the Nature and Observability of the Causal Relationship." In *Causation*, edited by Ernest Sosa and Michael Tooley, 125–36. Oxford: Oxford University Press.

Dupré, John. 1993. *The Disorder of Things: Metaphysical Foundations of the Disunity of Science*. Cambridge, MA: Harvard University Press.

Dupré, John. 2008. *The Constituents of Life*. Assen: Koninklijke Van Gorcum.

Dupré, John. 2012. *Processes of Life: Essays in the Philosophy of Biology*. Oxford: Oxford University Press.

Dupré, John. 2013. "Living Causes." *Aristotelian Society Supplementary Volume* 87 (1): 19–37. doi:10.1111/j.1467-8349.2013.00218.x.

Dupré, John, and Maureen A. O'Malley. 2007. "Metagenomics and Biological Ontology." *Studies in History and Philosophy of Biological and Biomedical Sciences* 38 (4): 834–46. doi:10.1016/j.shpsc.2007.09.001.

Eliasmith, Chris, and Oliver Trujillo. 2014. "The Use and Abuse of Large-Scale Brain Models." *Current Opinion in Neurobiology* 25: 1–6. doi:10.1016/j.conb.2013.09.009.

Ellis, Brian. 2001. *Scientific Essentialism*. Cambridge: Cambridge University Press.

Elster, Jon. 1989. *Nuts and Bolts for the Social Sciences*. Cambridge: Cambridge University Press.

Elster, Jon. 2007. Explaining Social Behaviour: More Nuts and Bolts for the Social Sciences. Revised Edition. Cambridge: Cambridge University Press.

Endicott, Ronald P. 2011. "Flat versus Dimensioned: The What and the How of Functional Realization." *Journal of Philosophical Research* 36: 191–208.

Felline, Laura. 2015. "Mechanisms Meet Structural Explanation." *Synthese*. Springer Netherlands. doi:10.1007/s11229-015-0746-9.

Fine, Kit. 2012. "Guide to Ground." In *Metaphysical Grounding: Understanding the Structure of Reality*, edited by Fabrice Correia and Benjamin Schnieder, 8–25. Cambridge: Cambridge University Press.

Fodor, Jerry. 1974. "Special Sciences (or: The Disunity of Science as a Working Hypothesis)." *Synthese* 28 (2): 97–115.

Forber, Patrick. 2010. "Confirmation and Explaining How Possible." *Studies in History and Philosophy of Science Part C: Studies in History and Philosophy of Biological and Biomedical Sciences* 41 (1): 32–40. doi:10.1016/j.shpsc.2009.12.006.

Franklin-Hall, Laura R. 2016. "New Mechanistic Explanation and the Need for Explanatory Constraints." In *Scientific Composition and Metaphysical Ground*, edited by Kenneth Aizawa and Carl Gillett, 41–76. London: Palgrave Macmillan.

Friedman, Michael. 1974. "Explanation and Scientific Understanding." *Journal of Philosophy* 71 (1): 5–19.

Gaddis, John Lewis. 2002. *The Landscape of History: How Historians Map the Past*. New York: Oxford University Press.

Garson, Justin. 2013. "The Functional Sense of Mechanism." *Philosophy of Science* 80 (3): 317–33.

Gerstman, L.J. 1968. "Classification of Self-Normalized Vowels." *IEEE Transactions on Audio Electroacoustics* AU-16: 78–80.

Gervais, Raoul, and Erik Weber. 2013. "Plausibility versus Richness in Mechanistic Models." *Philosophical Psychology* 26 (1): 139–52.

Giere, Ronald N. 1988. *Explaining Science: A Cognitive Approach*. Chicago: University of Chicago Press.

Giere, Ronald N. 1995. "The Skeptical Perspective: Science without Laws of Nature." In *Laws of Nature: Essays on the Philosophical, Scientific and Historical Dimensions*, edited by Friedel Weinert, 120–38. Berlin: Walter de Gruyter.

Giere, Ronald N. 1999. "Using Models to Represent Reality." In *Model-Based Reasoning in Scientific Discovery*, edited by L. Magnani, N. Nersessian, and P. Thagard, 41–57. New York: Springer.

Giere, Ronald N. 2004. "How Models Are Used to Represent Reality." *Philosophy of Science* 71 (5): 742–52.

Giere, Ronald N. 2006. *Scientific Perspectivism*. Chicago: University of Chicago Press.

Gillett, Carl. 2003. "The Metaphysics of Realization, Multiple Realizability, and the Special Sciences." Journal of Philosophy 100 (11): 591–603.

Gillett, Carl. 2007. "Understanding the New Reductionism: The Metaphysics of Science and Compositional Reduction." *Journal of Philosophy* 104 (4): 193–216.

Glennan, Stuart. 1992. "Mechanisms, Models and Causation." PhD dissertation. The University of Chicago.

Glennan, Stuart. 1996. "Mechanisms and the Nature of Causation." *Erkenntnis* 44 (1): 49–71.

Glennan, Stuart. 1997a. "Probable Causes and the Distinction between Subjective and Objective Chance." *Nous* 31 (4): 496–519. doi:10.1111/0029-4624.00058.

Glennan, Stuart. 1997b. "Capacities, Universality, and Singularity." *Philosophy of Science* 64 (4): 605–26. doi:10.1086/392574.

Glennan, Stuart. 2002a. "Rethinking Mechanistic Explanation." *Philosophy of Science* 69 (S3): S342–53. doi:10.1086/341857.

Glennan, Stuart. 2002b. "Contextual Unanimity and the Units of Selection Problem." *Philosophy of Science* 69 (1): 118–37. doi:10.1086/338944.

Glennan, Stuart. 2005. "Modeling Mechanisms." *Studies in the History of the Biological and Biomedical Sciences* 36 (2): 443–64. doi:10.1016/j.shpsc.2005.03.011.

Glennan, Stuart. 2008. "Mechanisms." In *The Routledge Companion to the Philosophy of Science*, edited by Stathis Psillos and Martin Curd, 376–84. London: Routledge.

Glennan, Stuart. 2009. "Productivity, Relevance and Natural Selection." *Biology & Philosophy* 24 (3): 325–39. doi:10.1007/s10539-008-9137-7.

Glennan, Stuart. 2010a. "Mechanisms, Causes, and the Layered Model of the World." *Philosophy and Phenomenological Research* 81 (2): 362–81. doi:10.1111/j.1933-1592.2010.00375.x.

Glennan, Stuart. 2010b. "Ephemeral Mechanisms and Historical Explanation." *Erkenntnis* 72 (2): 251–66. doi:10.1007/s10670-009-9203-9.

Glennan, Stuart. 2011. "Singular and General Causal Relations: A Mechanist Perspective." In *Causality in the Sciences*, edited by Phyllis McKay Illari, Federica Russo, and Jon Williamson, 789–817. Oxford: Oxford University Press.

Glennan, Stuart. 2014. "Aspects of Human Historiographic Explanation: A View from the Philosophy of Science." In *Explanation in the Special Sciences*, edited by Marie I. Kaiser, Oliver R. Scholz, Daniel Plenge, and Andreas Hüttemann, 273–91. Dordrecht: Springer Netherlands. doi:10.1007/978-94-007-7563-3.

Glennan, Stuart. 2015a. "When Is It Mental?" *Humana Mente* 29: 141–66. doi:10.2307/2408229.

Glennan, Stuart. 2015b. "Mechanisms and Mechanical Philosophy." In *Oxford Companion to the Philosophy of Science*, edited by Paul Humphreys, 1–17. doi:10.1093/oxfordhb/9780199368815.013.39.

Glennan, Stuart, and Phyllis Illari, eds. 2018. *The Routledge Handbook of Mechanisms and Mechanical Philosophy*. London: Routledge.

Godfrey-Smith, Peter. 2007. "Conditions for Evolution by Natural Selection." *The Journal of Philosophy* 104 (10): 489–516.

Godfrey-Smith, Peter. 2009a. *Darwinian Populations and Natural Selection*. Oxford: Oxford University Press.

Godfrey-Smith, Peter. 2009b. "Causal Pluralism." In *The Oxford Handbook of Causation*, edited by Helen Beebee, Christopher Hitchcock, and Peter Menzies, 326–37. New York: Oxford University Press.

Goodman, Nelson. 1955. *Fact, Fiction and Forecast*. Cambridge, MA: Harvard University Press.

Goodwin, William. 2018. "The Origins of the Reaction Mechanism." In *The Routledge Handbook of Mechanisms and Mechanical Philosophy*, edited by Stuart Glennan and Phyllis Illari, 46–58. London: Routledge.

Gould, Stephen Jay. 1990. *Wonderful Life: The Burgess Shale and the Nature of History*. 1st ed. New York: Norton.

Gould, Stephen Jay, and Richard C. Lewontin. 1979. "The Spandrels of San Marco and the Panglossian Paradigm: A Critique of the Adaptationist Programme." *Proceedings of the Royal Society of London. Series B. Biological Sciences* 205 (1161): 581–98.

Griffiths, Paul, and Russell Gray. 1994. "Developmental Systems and Evolutionary Explanation." *The Journal of Philosophy* 91 (6): 277–304.

Grimm, V., E. Revilla, Uta Berger, F. Jeltsch, W. Mooij, S. Railsback, H. Thulke, J. Weiner, T. Wiegand, and Donald L. DeAngelis. 2005. "Pattern-Oriented Modeling of Agent-Based Complex Systems: Lessons from Ecology." *Science* 310 (5750): 987–91.

Gross, Fridolin. 2015. "The Relevance of Irrelevance: Explanation in Systems Biology." In *Explanation in Biology: An Enquiry into the Diversity of Explanatory Patterns in the Life Sciences*, edited by Pierre-Alain Braillard and Christophe Malaterre, 175–98. New York: Springer.

Grush, Rick. 2003. "In Defense of Some 'Cartesian' Assumptions Concerning the Brain and Its Operation." *Biology and Philosophy* 18 (1): 53–93.

Hacking, Ian. 1986. "Making Up People." In *Reconstructing Individualism: Autonomy, Individuality, and the Self in Western Thought*, edited by Thomas Heller, David E Wellbery, and Morton Sosna, 222–36. Stanford, CA: Stanford University Press.

Hacking, Ian. 1999. *The Social Construction of What?* Cambridge, MA: Harvard University Press.

Hall, Ned. 2004. "Two Concepts of Causation." In *Causation and Counterfactuals*, edited by John Collins, Ned Hall, and L. A. Paul, 225–76. Cambridge, MA: Bradford Book/MIT Press.

Handfield, Toby, Charles R. Twardy, Kevin B. Korb, and Graham Oppy. 2008. "The Metaphysics of Causal Models: Where's the Biff?" *Erkenntnis* 68 (2): 149–68.

Harbecke, Jens. 2010. "Mechanistic Constitution in Neurobiological Explanations." *International Studies in the Philosophy of Science* 24 (3): 267–85.

Haugeland, John. 2000. "Mind Embodied and Embedded." In *Having Thought: Essays in the Metaphysics of Mind*, 207–37. Cambridge, MA: Harvard University Press.

Hausman, Daniel M., and James Woodward. 1999. "Independence, Invariance and the Causal Markov Condition." *British Journal for the Philosophy of Science* 50: 521–83. doi:10.1093/bjps/50.4.521.

Hausman, Daniel M., and James Woodward. 2004. "Manipulation and the Causal Markov Condition." *Philosophy of Science* 71 (5): 846–56.

Hedström, Peter, and Richard Swedberg. 1996. "Social Mechanisms." *Acta Sociologica* 39: 281–308.

Hedström, Peter, and Richard Swedberg. 1998. "Social Mechanisms: An Introductory Essay." In *Social Mechanisms: An Analytical Approach to Social Theory*, edited by Peter Hedström and Richard Swedberg, 1–31. Cambridge, UK; New York: Cambridge University Press.

Hedström, Peter, and Petri Ylikoski. 2010. "Causal Mechanisms in the Social Sciences." *Annual Review of Sociology* 36 (1): 49–67. doi:10.1146/annurev.soc.012809.102632.

Heil, John. 2003. *From an Ontological Point of View*. Oxford: Clarendon Press.

Heil, John. 2012. *The Universe as We Find It*. Oxford: Clarendon Press.

Hempel, Carl G. 1965. *Aspects of Scientific Explanation, and Other Essays in the Philosophy of Science*. New York: Free Press.

Hempel, Carl G. 1966. *Philosophy of Natural Science*. Englewood Cliffs, NJ: Prentice-Hall.

Hempel, Carl G, and Paul Oppenheim. 1948. "Studies in the Logic of Explanation." *Philosophy of Science* 15 (2): 135–75. doi:10.1086/287002.

Hitchcock, Christopher Read. 1995. "Discussion: Salmon on Explanatory Relevance." *Philosophy of Science* 62 (2): 304–20.

Hochstein, Eric. 2015. "One Mechanism, Many Models: A Distributed Theory of Mechanistic Explanation." *Synthese* 193 (5): 1387–407. doi:10.1007/s11229-015-0844-8.

Hodgkin, Alan L., and Andrew F. Huxley. 1939. "Action Potentials Recorded from Inside a Nerve Fibre." *Nature* 144: 710–11.

Hodgkin, Alan L., and Andrew F. Huxley. 1952. "A Quantitative Description of Membrane Current and Its Application to Conduction and Excitation in Nerve." *The Journal of Physiology*, 500–44.

Hoefer, Carl. 2007. "The Third Way on Objective Probability: A Sceptic's Guide to Objective Chance." *Mind* 116 (463): 549–96. doi:10.1093/mind/fzm549.

Huffaker, C. 1958. "Experimental Studies on Predation: Dispersion Factors and Predator-Prey Oscillations." *Hilgardia: A Journal of the California Agricultural Experiment Station* 27 (14): 343–83.

Hughes, R.I.G. 1989. *The Structure and Interpretation of Quantum Mechanics*. Cambridge, MA: Harvard University Press.

Hull, David L., Rodney E. Langman, and Sigrid S. Glenn. 2001. "A General Account of Selection: Biology, Immunology, and Behavior." *Behavioral and Brain Sciences* 24 (3): 511–28.

Humphreys, Paul. 1989. *The Chances of Explanation*. Princeton: Princeton University Press.

Huneman, Philippe. 2010. "Topological Explanations and Robustness in Biological Sciences." *Synthese* 177 (2): 213–45.

Huneman, Philippe. 2015. "Diversifying the Picture of Explanations in Biological Sciences: Ways of Combining Topology with Mechanisms." *Synthese*. doi:10.1007/s11229-015-0808-z.

Hüttemann, Andreas. 2010. "Physicalism and the Part-Whole Relation." Working paper. LSE Centre for Philosophy of Natural and Social Science. http://www.lse.ac.uk/CPNSS/research/concludedResearchProjects/orderProject/documents/Publications/HuttemanPhysicalism.pdf.

Hüttemann, Andreas, and David Papineau. 2005. "Physicalism Decomposed." *Analysis* 65 (285): 33–9.

Illari, Phyllis. 2013. "Mechanistic Explanation: Integrating the Ontic and Epistemic." *Erkenntnis* 78 (July): 237–55. doi:10.1007/s10670-013-9511-y.

Illari, Phyllis, and Federica Russo. 2013. *Causality: Philosophical Theory Meets Scientific Practice*. Oxford: Clarendon Press.

Illari, Phyllis, and Jon Williamson. 2012. "What Is a Mechanism? Thinking about Mechanisms across the Sciences." *European Journal for Philosophy of Science* 2 (1): 119. doi:10.1007/s13194-011-0038-2.

Kaiser, Marie I., and Beate Krickel. 2016. "The Metaphysics of Constitutive Mechanistic Phenomena." *British Journal for the Philosophy of Science* 0: 1–35. doi:10.1093/bjps/axv058.

Kaplan, David Michael. 2011. "Explanation and Description in Computational Neuroscience." *Synthese* 183 (3): 339–73. doi:10.1007/s11229-011-9970-0.

Kaplan, David Michael. 2018. "Mechanisms and Dynamical Explanation." In *The Routledge Handbook of Mechanisms and Mechanical Philosophy*, edited by Stuart Glennan and Phyllis Illari, 267–80. London: Routledge.

Kaplan, David Michael, and William Bechtel. 2011. "Dynamical Models: An Alternative or Complement to Mechanistic Explanations?" *Topics in Cognitive Science* 3 (2): 438–44.

Kaplan, David Michael, and Carl F. Craver. 2011. "The Explanatory Force of Dynamical and Mathematical Models in Neuroscience: A Mechanistic Perspective." *Philosophy of Science* 78 (4): 601–27.

Kauffman, Stuart A. 1970. "Articulation of Parts Explanation in Biology and the Rational Search for Them." In *PSA: Proceedings of the Biennial Meeting of the Philosophy of Science Association*, 257–72. Boston: Philosophy of Science Association.

Kauffman, Stuart A. 1993. *The Origins of Order: Self-Organization and Selection in Evolution*. New York: Oxford University Press.

Kauffman, Stuart A. 1995. *At Home in the Universe: The Search for Laws of Self-Organization and Complexity*. New York: Oxford University Press.

Keller, Evelyn Fox. 2008. "Organisms, Machines, and Thunderstorms: A History of Self-Organization, Part One." *Historical Studies in the Natural Sciences* 38 (1): 45–75. doi:10.1525/hsns.2008.38.1.45.

Keller, Evelyn Fox. 2009. "Organisms, Machines, and Thunderstorms: A History of Self-Organization, Part Two. Complexity, Emergence, and Stable Attractors." *Historical Studies in the Natural Sciences* 39 (1): 1–31. doi:10.1525/hsns.2009.39.1.1.

Kellert, Stephen H., Helen E. Longino, and C. Kenneth Waters. 2006. "The Pluralist Stance." In *Scientific Pluralism*, edited by Stephen H. Kellert, Helen E. Longino, and C. Kenneth Waters. Minnesota Studies in the Philosophy of Science, vol. 19. Minneapolis: University of Minnesota Press.

Khalidi, M. A. 2013. "Kinds (Natural Kinds vs. Human Kinds)." In *Encyclopedia of Philosophy and the Social Sciences*, edited by Brian Kaldis. Thousand Oaks, CA: Sage.

Kim, Jaegwon. 1973. "Causation, Nomic Subsumption, and the Concept of Event." *Journal of Philosophy* 70 (8): 217–36.

Kim, Jaegwon. 1974. "Noncausal Connections." *Noûs* 8 (1): 41–52.

Kim, Jaegwon. 1976. "Events as Property Exemplifications." In *Action Theory*, edited by Miles Brand and D Walton, 159–77. Dordrecht: D. Reidel.

Kim, Jaegwon. 1984. "Concepts of Supervenience." *Philosophy and Phenomenological Research* 45 (December): 153–76.

Kistler, Max. 2009. "Mechanisms and Downward Causation." *Philosophical Psychology* 6 (5): 595–609.

Kitcher, Philip. 1984. "1953 and All That: A Tale of Two Sciences." *Philosophical Review* 93: 335–74.

Kitcher, Philip. 1989. "Explanatory Unification and the Causal Structure of the World." In *Scientific Explanation*, edited by Philip Kitcher and Wesley C. Salmon, 410–506. Minnesota Studies in the Philosophy of Science, vol. 13. Minneapolis: University of Minnesota Press.

Kuhlmann, Meinard. 2011. "Mechanisms in Dynamically Complex Systems." In *Causality in the Sciences*, edited by Phyllis McKsay Illari, Federica Russo, and Jon Williamson, 880–906. Oxford: Oxford University Press.

Kuhlmann, Meinard. 2015. "A Mechanistic Reading of Quantum Laser Theory." In *Why Is More Different? Philosophical Issues in Condensed Matter Physics and Complex Systems*, edited by Brigitte Falkenburg and Margaret Morrison, 251–71. Berlin: Springer.

Kuhlmann, Meinard. 2018. "Mechanisms in Physics." In *The Routledge Handbook of Mechanisms and Mechanical Philosophy*, edited by Stuart Glennan and Phyllis Illari, 283–95. London: Routledge.

Kuhlmann, Meinard, and Stuart Glennan. 2014. "On the Relation between Quantum Mechanical and Neo-Mechanistic Ontologies and Explanatory Strategies." *European Journal for Philosophy of Science* 4 (3): 337–59. doi:10.1007/s13194-014-0088-3.

Kuhn, Thomas S. 1957. *The Copernican Revolution: Planetary Astronomy in the Development of Western Thought*. Cambridge, MA: Harvard University Press.

Kuhn, Thomas S. 1991. "The Natural and the Human Sciences." In *The Interpretive Turn: Philosophy, Science, Culture*, edited by David R. Hiley, James F. Bohman, and Richard Shusterman, 17–24. Ithaca, NY: Cornell University Press.

Ladyman, James. 2014. "Structural Realism." In *Stanford Encyclopedia of Philosophy* (Spring 2014 edition), edited by Edward N. Zalta.

Ladyman, James, and Don Ross. 2007. *Every Thing Must Go: Metaphysics Naturalized*. Oxford: Oxford University Press.

Lange, Marc. 2013. "What Makes a Scientific Explanation Distinctively Mathematical?" *British Journal for the Philosophy of Science* 64: 485–511. doi:10.1093/bjps/axs012.

Lange, Marc. 2015. "On 'Minimal Model Explanations': A Reply to Batterman and Rice." *Philosophy of Science* 82 (2): 292–305. doi:10.1086/680488.

Leuridan, Bert. 2012. "Three Problems for the Mutual Manipulability Account of Constitutive Relevance in Mechanisms." *British Journal for the Philosophy of Science* 63 (2): 399–427. doi:10.1093/bjps/axr036.

Levin, Simon A. 1992. "The Problem of Pattern and Scale in Ecology." *Ecology* 73 (6): 1943–67.

Levy, Arnon. 2013. "Three Kinds of New Mechanism." *Biology & Philosophy* 28: 99–114.

Levy, Arnon, and William Bechtel. 2013. "Abstraction and the Organization of Mechanisms." *Philosophy of Science* 80 (2): 241–61.

Lewis, David K. 1972. "Psychophysical and Theoretical Identifications." *Australasian Journal of Philosophy* 50 (December): 249–58.

Lewis, David K. 1973. "Causation." *The Journal of Philosophy* 70: 556–67.

Lewis, David K. 1979. "Counterfactual Dependence and Time's Arrow." *Noûs* 13: 455–76.

Lewis, David K. 2004. "Void and Object." In *Causation and Counterfactuals*, edited by John Collins, Ned Hall, and L. A. Paul, 277–90. Cambridge, MA: Bradford Book/MIT Press.

Lewontin, Richard C. 1970. "The Units of Selection." *Annual Review of Ecology and Systematics* 1: 1–18.

Little, Daniel. 2006. "Levels of the Social." In *Philosophy of Anthropology and Sociology*, edited by Stephen Turner and Mark Risjord, 343–71. Amsterdam and New York: Elsevier.

Little, Daniel. 2011. "Causal Mechanisms in the Social Realm." In *Causality in the Sciences*, edited by Phyllis McKay Illari, Federica Russo, and Jon Williamson, 273–95. Oxford: Oxford University Press.

Little, Daniel. 2018. "Disaggregating Historical Explanation: The Move to Social Mechanisms." In *The Routledge Handbook of Mechanisms and Mechanical Philosophy*, edited by Stuart Glennan and Phyllis Illari, 413–22. London: Routledge.

Lloyd, Elizabeth. 2010. "Confirmation and Robustness of Climate Models." *Philosophy of Science* 77 (5): 971–84.

Love, Alan C. 2008. "Typology Reconfigured: From the Metaphysics of Essentialism to the Epistemology of Representation." *Acta Biotheoretica* 57 (1–2): 51–75.

Love, Alan C. 2011. "Hierarchy, Causation and Explanation: Ubiquity, Locality and Pluralism." *Interface Focus* 2 (1): 115–25. doi:10.1098/rsfs.2011.0064.

Machamer, Peter. 2004. "Activities and Causation: The Metaphysics and Epistemology of Mechanisms." *International Studies in the Philosophy of Science* 18 (1): 27–39.

Machamer, Peter, Lindley Darden, and Carl F. Craver. 2000. "Thinking about Mechanisms." *Philosophy of Science* 67 (1): 1–25.

Mackie, J.L. 1965. "Causes and Conditions." *American Philosophical Quarterly* 2 (4): 245–64.

Mahoney, James. 2001. "Beyond Correlational Analysis: Recent Innovations in Theory and Method." *Sociological Forum* 16 (3): 575–93.

Matthen, Mohan, and André Ariew. 2002. "Two Ways of Thinking about Fitness and Natural Selection." *Journal of Philosophy* 99 (2): 55–83.

Mayr, Ernst. 1959. "Typological versus Population Thinking." In *Evolution and Anthropology: A Centennial Appraisal*, edited by B. J. Meggers, 409–12. Washington, DC: Anthropological Society of Washington.

Mayr, Ernst. 1961. "Cause and Effect in Biology." *Science* 134 (November): 1501–6.

McGee, Harold. 2004. *On Food and Cooking: The Science and Lore of the Kitchen*. Kindle Edition. New York: Scribner.

McMullin, Ernan. 1985. "Galilean Idealization." *Studies in History and Philosophy of Science* 16: 247–73.

Menzies, Peter. 1996. "Probabilistic Causation and the Pre-Emption Problem." *Mind* 105 (417): 85–117.

Menzies, Peter. 1999. "Intrinsic versus Extrinsic Conceptions of Causation." In *Causation and Laws of Nature*, edited by Howard Sankey, 313–29. Dordrecht: Kluwer.

Menzies, Peter. 2009. "Platitudes and Counterexamples." In *The Oxford Handbook of Causation*, edited by Helen Beebee, Christopher Hitchcock, and Peter Menzies, 341–67. New York: Oxford University Press.

Menzies, Peter, and Huw Price. 1993. "Causation as a Secondary Quality." *The British Journal for the Philosophy of Science* 44 (2): 187–203.

Mesoudi, Alex, Andrew Whiten, and Kevin Laland. 2006. "Towards a Unified Science of Cultural Evolution." *The Behavioral and Brain Sciences* 29 (4): 1–55. doi:10.1017/S0140525X06009083.

Miller, Erin. 2015. "Overview: Terrorism in 2014." College Park, MD. http://www.start.umd.edu/pubs/START_GTD_OverviewofTerrorism2014_Aug2015.pdf.

Millikan, Ruth Garrett. 1984. *Language, Thought, and Other Biological Categories*. Cambridge, MA: Bradford Book/MIT Press.

Millstein, Roberta L. 2009. "Populations as Individuals." *Biological Theory* 4 (3): 267–73.

Millstein, Roberta L. 2013. "Natural Selection and Causal Productivity." In *Mechanism and Causality in Biology and Economics*, edited by Hsiang-Ke Chao, Szu-Ting Chen, and L. Roberta Millstein, 147–63. Dordrecht: Springer Netherlands. doi:10.1007/978-94-007-2454-9_8.

Mitchell, Sandra D. 2009. *Unsimple Truths: Science, Complexity and Policy*. Chicago: University of Chicago Press.

Moreno, Alvaro, and Kepa Ruiz-Mirazo. 2009. "The Problem of the Emergence of Functional Diversity in Prebiotic Evolution." *Biology and Philosophy* 24 (5): 585–605.

Morgan, Mary S., and Margaret Morrison. 1999. *Models as Mediators: Perspectives on Natural and Social Science.* Cambridge: Cambridge University Press.

Morrison, Margaret. 2007. "Where Have All the Theories Gone?" *Philosophy of Science* 74 (2): 195–228.

Mossio, Matteo, Cristian Saborido, and Alvaro Moreno. 2009. "An Organizational Account of Biological Functions." *The British Journal for the Philosophy of Science* 60 (4): 1–23.

Müller, Gerd B., and Stuart A. Newman. 2005. "The Innovation Triad: An EvoDevo Agenda." *Journal of Experimental Zoology Part B: Molecular and Developmental Evolution* 304B (6): 487–503.

Mulroney, Susan E., Adam K. Myers, and Frank H. Netter. 2016. *Netter's Essential Physiology.* 2nd ed. Philadelphia, PA: Elsevier.

Nagel, Ernest. 1979. *The Structure of Science: Problems in the Logic of Scientific Explanation.* 2nd ed. Indianapolis, IN: Hackett.

Nature Chemical Biology. 2007. "A Mechanistic Meeting Point." *Nature Chemical Biology* 3 (3): 127.

Neander, Karen. 1991. "Functions as Selected Effects: The Conceptual Analyst's Defense." *Philosophy of Science* 58 (2): 168–84.

Nicholson, Daniel J. 2012. "The Concept of Mechanism in Biology." *Studies in History and Philosophy of Science Part C: Studies in History and Philosophy of Biological and Biomedical Sciences* 43 (1): 152–63.

Nicholson, Daniel J. 2013. "Organisms ≠ Machines." *Studies in History and Philosophy of Science Part C: Studies in History and Philosophy of Biological and Biomedical Sciences* 44 (4): 669–78.

Noble, Denis. 2013. "Physiology Is Rocking the Foundations of Evolutionary Biology." *Experimental Physiology* 98 (8): 1235–43.

O'Brien Jr, James M., Naeem A. Ali, Scott K. Aberegg, and Edward Abraham. 2007. "Sepsis." *The American Journal of Medicine* 120 (12): 1012–22. doi:10.1016/j.amjmed.2007.01.035.

Oyama, Susan. 1985. *The Ontogeny of Information: Developmental Systems and Evolution.* Cambridge: Cambridge University Press.

Papineau, David. 2008. "Must a Physicalist Be a Microphysicalist." In *Being Reduced: New Essays on Reduction, Explanation, and Causation,* edited by Jakob Hohwy and Jesper Kallestrup, 126–48. Oxford: Oxford University Press.

Pettit, Philip. 1993. "A Definition of Physicalism." *Analysis* 53 (4): 213–23.

Piccinini, Gualtiero. 2007a. "Computing Mechanisms." *Philosophy of Science* 74 (4): 501–26.

Piccinini, Gualtiero. 2007b. "Computational Modelling vs. Computational Explanation: Is Everything a Turing Machine, and Does It Matter to the Philosophy of Mind?" *Australasian Journal of Philosophy* 85 (1): 93–115. doi:10.1080/00048400601176494.

Piccinini, Gualtiero. 2008. "Computers." *Pacific Philosophical Quarterly* 89 (March): 32–73.

Piccinini, Gualtiero. 2010. "The Mind as Neural Software? Understanding Functionalism, Computationalism, and Computational Functionalism." *Philosophy and Phenomenological Research* 81 (2): 269–311.

Piccinini, Gualtiero, and Sonya Bahar. 2013. "Neural Computation and the Computational Theory of Cognition." *Cognitive Science* 37 (3): 453–88. doi:10.1111/cogs.12012.

Piccinini, Gualtiero, and Carl Craver. 2011. "Integrating Psychology and Neuroscience: Functional Analyses as Mechanism Sketches." *Synthese* 183(3): 283–311. https://doi.org/10.1007/s11229-011-9898-4.

Plutynski, Anya. 2013. "Cancer and the Goals of Integration." *Studies in History and Philosophy of Science Part C: Studies in History and Philosophy of Biological and Biomedical Sciences* 44 (4): 466–76. doi:10.1016/j.shpsc.2013.03.019.

Pollan, Michael. 2009. *In Defense of Food*. New York: Penguin.

Popa, Tiberiu. 2018. "Mechanisms: Ancient Sources." In *The Routledge Handbook of Mechanisms and Mechanical Philosophy*, edited by Stuart Glennan and Phyllis Illari, 13–25. London: Routledge.

Povich, Mark. 2016. "Minimal Models and the Generalized Ontic Conception of Scientific Explanation." *British Journal for the Philosophy of Science*. doi: 10.1093/bjps/axw019.

Povich, Mark, and Carl F. Craver. 2018. "Mechanistic Levels, Reduction, and Emergence." In *The Routledge Handbook of Mechanisms and Mechanical Philosophy*, edited by Stuart Glennan and Phyllis Illari, 185–97. London: Routledge.

Psillos, Stathis. 2004. "A Glimpse of the Secret Connexion: Harmonizing Mechanisms with Counterfactuals." *Perspectives on Science* 12 (3): 288–319.

Psillos, Stathis. 2009. "Regularity Theories." In *The Oxford Handbook of Causation*, edited by Helen Beebee, Christopher Hitchcock, and Peter Menzies, 131–57. New York: Oxford University Press.

Psillos, Stathis. 2011. "Choosing the Realist Framework." *Synthese* 180 (2): 301–16. doi:10.1007/s11229-009-9606-9.

Putnam, Hilary. 1973. "Reductionism and the Nature of Psychology." *Cognition: International Journal of Cognitive Science* 2: 131–46.

Quine, Willard V. 1951. "Two Dogmas of Empiricism." *Philosophical Review* 60 (1): 20–43.

Quine, Willard V. 1969. "Natural Kinds." In *Essays in Honor of Carl G. Hempel*, edited by Nicholas Rescher, 5–24. Dordrecht: Reidel.

Rescher, Nicholas. 1996. *Process Metaphysics: An Introduction to Process Philosophy*. Albany: State University of New York Press.

Resnik, David. 1991. "How-Possibly Explanations in Biology." *Acta Biotheoretica* 39 (2): 141–9.

Reydon, Thomas. 2012. "How-Possibly Explanations as Genuine Explanations and Helpful Heuristics: A Comment on Forber." *Studies in History and Philosophy of Science Part C: Studies in History and Philosophy of Biological and Biomedical Sciences* 43 (1): 302–10.

Richardson, Robert C. 2001. "Complexity, Self-Organization and Selection." *Biology and Philosophy* 16 (5): 653–82.

Rosen, Gideon. 2010. "Metaphysical Dependence: Grounding and Reduction." In *Modality: Metaphysics, Logic, and Epistemology*, edited by Bob Hale and Aviv Hoffmann, 109–35. Oxford: Oxford University Press.

Roux, Sophie. 2018. "The Rise of Mechanical Philosophy." In *The Routledge Handbook of Mechanisms and Mechanical Philosophy*, edited by Stuart Glennan and Phyllis Illari, 26–45. London: Routledge.

Russo, Federica, and Jon Williamson. 2007. "Interpreting Causality in the Health Sciences." *International Studies in the Philosophy of Science* 21 (2): 157–70. doi:10.1080/02698590701498084.

Salmon, Wesley C. 1981. "Causality: Production and Propagation." *PSA 1980: Proceedings of the Biennial Meeting of the Philosophy of Science Association,* vol. 2: 49–69. East Lansing, MI: Philosophy of Science Association.

Salmon, Wesley C. 1984. *Scientific Explanation and the Causal Structure of the World.* Princeton: Princeton University Press.

Salmon, Wesley C. 1985. "Scientific Explanation: Three Basic Conceptions." In *PSA 1984: Proceedings of the 1984 Biennial Meetings of the Philosophy of Science Association,* vol. 2: 293–305. East Lansing, MI: Philosophy of Science Association.

Salmon, Wesley C. 1989. "Four Decades of Scientific Explanation." In *Scientific Explanation,* edited by Philip Kitcher and Wesley C. Salmon, 3–219. Minnesota Studies in Philosophy of Science, vol. 13. Minneapolis: University of Minnesota Press.

Salmon, Wesley C. 1994. "Causality without Counterfactuals." *Philosophy of Science* 61 (2): 297–312.

Salmon, Wesley C. 1997. "Causality and Explanation: A Reply to Two Critiques." *Philosophy of Science* 64: 461–77.

Sanford, David. 2014. "Determinates vs. Determinables." In *Stanford Encyclopedia of Philosophy* (Winter 2014 edition), edited by Edward N. Zalta.

Schaffer, Jonathan. 2000. "Causation by Disconnection." *Philosophy of Science* 67 (2): 285–300.

Schaffer, Jonathan. 2003. "Is There a Fundamental Level?" *Noûs* 37 (3): 498–517.

Schaffer, Jonathan. 2009a. "On What Grounds What." In *Metametaphysics,* edited by David Chalmers, David Manley, and Ryan Wasserman, 347–83. Oxford: Oxford University Press.

Schaffer, Jonathan. 2009b. "Monism: The Priority of the Whole." *Philosophical Review* 119 (1): 31–76. doi:10.1215/00318108-2009-025.

Schaffer, Jonathan. 2014. "The Metaphysics of Causation." In *Stanford Encyclopedia of Philosophy* (Spring 2014 edition), edited by Edward N. Zalta.

Sebesta, Robert W. 1999. *Concepts of Programming Languages.* 4th ed. Reading, MA: Addison-Wesley.

Sellars, Wilfred. 1963. "Philosophy and the Scientific Image of Man." In *Empiricism and the Philosophy of Mind,* 1–40. London: Routledge & Kegan Paul.

Shapin, Steven. 1996. *The Scientific Revolution.* Chicago: University of Chicago Press.

Sider, Theodore. 2007. "Parthood." *Philosophical Review* 116 (1): 51–91. doi:10.1215/00318108-2006-022.

Simon, Herbert A. 1996. *The Sciences of the Artificial.* 3rd ed. Cambridge, MA: MIT Press.

Skipper, Robert A. Jr, and Roberta L. Millstein. 2005. "Thinking about Evolutionary Mechanisms: Natural Selection." *Studies in History and Philosophy of Science Part C: Studies in History and Philosophy of Biological and Biomedical Sciences* 36 (2): 327–47.

Sober, Elliott. 1983. "Equilibrium Explanation." *Philosophical Studies* 43 (2): 201–10. doi:10.1007/BF00372383.

Sober, Elliott. 2010. "Evolutionary Theory and the Reality of Macro-Probabilities." In *The Place of Probability in Science: In Honor of Ellery Eells (1953–2006),* edited by Ellery Eells and J. H. Fetzer, 133–61. Dordrecht: Springer Netherlands. doi:10.1007/978-90-481-3615-5_6.

Steel, Daniel. 2005. "Mechanisms and Functional Hypotheses in Social Science." *Philosophy of Science* 72: 941–52. doi:10.1086/508951.

Steel, Daniel. 2006. "Comment on Hausman & Woodward on the Causal Markov Condition." *British Journal for the Philosophy of Science.* doi:10.1093/bjps/axi154.

Steel, Daniel. 2008. *Across the Boundaries: Extrapolation in Biology and Social Science*. New York: Oxford University Press.

Sterelny, Kim. 1996. "Explanatory Pluralism in Evolutionary Biology." *Biology and Philosophy* 11: 193–214.

Stinson, Catherine. 2016. "Mechanisms in Psychology: Ripping Nature at Its Seams." *Synthese* 193. Springer Netherlands: 1585–614. doi:10.1007/s11229-015-0871-5.

Strevens, Michael. 2005. "How Are the Sciences of Complex Systems Possible?" *Philosophy of Science* 72 (4): 531–56. doi:10.1086/505471.

Strevens, Michael. 2008. *Depth: An Account of Scientific Explanation*. Cambridge, MA: Harvard University Press.

Strevens, Michael. 2011. "Probability out of Determinism." In *Probabilities in Physics*, edited by Claus Beisbart and Stephan Hartmann, 339–64. Oxford: Oxford University Press. doi:10.1093/acprof:oso/9780199577439.003.0013.

Strevens, Michael. 2013. "Causality Reunified." *Erkenntnis* 78 (Supplement 2): 299–320.

Suppe, Frederick. 1977. *The Structure of Scientific Theories*. Chicago: University of Illinois Press.

Tabery, James G. 2004. "Synthesizing Activities and Interactions in the Concept of a Mechanism." *Philosophy of Science* 71 (1): 1–15.

Taylor, Charles. 1971. "Interpretation and the Sciences of Man." *The Review of Metaphysics* 25 (1): 3–51.

Teller, Paul. 2001. "Twilight of the Perfect Model Model." *Erkenntnis* 55 (3): 393–415.

Thagard, Paul. 1999. *How Scientists Explain Disease*. Princeton, NJ: Princeton University Press.

Thagard, Paul, and Fred Kroon. 2006. *Hot Thought: Mechanisms and Applications of Emotional Cognition*. Cambridge, MA: MIT Press.

Thagard, Paul, and Josef Nerb. 2002. "Emotional Gestalts: Appraisal, Change, and the Dynamics of Affect." *Personality and Social Psychology Review* 6 (4): 274–82.

Tooley, Michael. 2009. "Causes, Laws, and Ontology." In *The Oxford Handbook of Causation*, edited by Helen Beebee, Christopher Hitchcock, and Peter Menzies, 368–86. Oxford: Oxford University Press.

Townsend, James T. 1992. "Don't Be Fazed by PHASER: Beginning Exploration of a Cyclical Motivational System." *Behavior Research Methods, Instruments, & Computers* 24 (2): 219–27. doi:10.3758/BF03203499.

van Fraassen, Bas C. 1980. "The Scientific Image," Oxford: Oxford University Press. doi:10.1093/0198244274.001.0001.

van Fraassen, Bas C. 2008. *Scientific Representation: Paradoxes of Perspective*. Oxford: Clarendon Press.

van Gelder, Tim. 1995. "What Might Cognition Be, If Not Computation?" *The Journal of Philosophy* 92 (7): 345–81.

Varela, Francisco, and Humberto Maturana. 1980. *Autopoiesis and Cognition: The Realization of the Living*. Dordrecht: Reidel.

von Bertalanffy, Ludwig. 1950. "An Outline of General System Theory." *British Journal for the Philosophy of Science* 1 (2): 134–65.

Wagar, Brandon, and Paul Thagard. 2004. "Spiking Phineas Gage: A Neurocomputational Theory of Cognitive-Affective Integration in Decision Making." *Psychological Review* 111 (1): 67–79.

Wagner, Günter P., Mihaela Pavlicev, and James M. Cheverud. 2007. "The Road to Modularity." *Nature Reviews. Genetics* 8: 921–31. doi:10.1038/nrg2267.

Waskan, Jonathan. 2011. "Mechanistic Explanation at the Limit." *Synthese* 183 (3): 389–408.

Wasserman, Ryan. 2015. "Material Constitution." In *Stanford Encyclopedia of Philosophy* (Spring 2015 edition), edited by Edward N. Zalta.

Waters, C. Kenneth. 2007. "Causes That Make a Difference." *The Journal of Philosophy* 104 (11): 551–79.

Watts, D. J., and S. H. Strogatz. 1998. "Collective Dynamics of 'Small-World' Networks." *Nature* 393: 440–2. doi:10.1038/30918.

Weber, Bruce H., and David J. Depew. 1996. "Natural Selection and Self-Organization." *Biology and Philosophy* 11 (1): 33–65.

Weber, Marcel. 2008. "Causes without Mechanisms: Experimental Regularities, Physical Laws, and Neuroscientific Explanation." *Philosophy of Science* 75 (5): 995–1007. doi:10.1086/594541.

Weisberg, Michael. 2007a. "Three Kinds of Idealization." *The Journal of Philosophy* 104 (12): 639–59.

Weisberg, Michael. 2007b. "Who Is a Modeler?" *The British Journal for the Philosophy of Science* 58 (2): 207–33. doi:10.1093/bjps/axm011.

Weisberg, Michael. 2013. *Simulation and Similarity: Using Models to Understand the World.* Oxford: Oxford University Press.

Westfall, Richard S. 1971. *The Construction of Modern Science: Mechanisms and Mechanics.* New York: Wiley.

Whitehead, Alfred N. 1929. *Process and Reality: An Essay in Cosmology.* New York: Macmillan.

Wiener, Norbert. 1950. *The Human Use of Human Beings: Cybernetics and Society.* Boston: Houghton Mifflin.

Wilensky, Uri. 1997. "NetLogo Wolf Sheep Predation Model." Center for Connected Learning and Computer-Based Modeling, Northwestern University, Evanston, IL.

Wilensky, Uri, and Kenneth Reisman. 2010. "Thinking Like a Wolf, a Sheep, or a Firefly: Learning Biology through Constructing and Testing Computational Theories—An Embodied Modeling Approach." *Cognition and Instruction*, August 2014: 37–41. doi:10.1207/s1532690xci2402.

Williamson, Jon. 2011. "Mechanistic Theories of Causality Part I." *Philosophy Compass* 6: 421–32.

Wilson, Jessica M. 2013. "A Determinable-Based Account of Metaphysical Indeterminacy." *Inquiry: An Interdisciplinary Journal of Philosophy* 56 (4): 359–85. doi:10.1080/0020174X.2013.816251.

Wimsatt, William C. 1972. "Complexity and Organization." In *PSA: Proceedings of the Biennial Meeting of the Philosophy of Science Association Vol. 1972*, edited by Kenneth Schaffner and Robert Cohen, 67–86. Dordrecht: D. Reidel.

Wimsatt, William C. 1980. "Reductionistic Research Strategies and Their Biases in the Units of Selection Controversy." In *Scientific Discovery: Case Studies*, edited by Thomas Nickles, 213–59. Dordrecht: Reidel.

Wimsatt, William C. 1994. "The Ontology of Complex Systems: Levels of Organization, Perspectives, and Causal Thickets." *Canadian Journal of Philosophy* Supplement: 207–74.

Wimsatt, William C. 2000. "Emergence as Non-Aggregativity and the Biases of Reductionisms." *Foundations of Science* 5 (3): 269–97.

Wimsatt, William C. 2001. "Generative Entrenchment and the Developmental Systems Approach to Evolutionary Process." In *Cycles of Contingency: Developmental Systems and Evolution*, edited by Russell D. Gray and Paul E. Griffiths, 219–38. Cambridge, MA: MIT Press.

Wimsatt, William C. 2002. "Using False Models to Elaborate Constraints on Processes: Blending Inheritance in Organic and Cultural Evolution." *Philosophy of Science* 69 (S3): S12–24. doi:10.1086/341764.

Wimsatt, William C. 2007. *Re-Engineering Philosophy for Limited Beings*. Cambridge, MA: Harvard University Press.

Winther, Rasmus Grønfeldt. 2009. "Part-Whole Science." *Synthese* 178 (3): 397–427.

Winther, Rasmus Grønfeldt. 2016. "The Structure of Scientific Theories." In *Stanford Encyclopedia of Philosophy* (Spring 2016 edition), edited by Edward N. Zalta.

Wise, M. Norton. 2011. "Science as (Historical) Narrative." *Erkenntnis* 75 (3): 349–76.

Wittgenstein, Ludwig. 1958. *Preliminary Studies for the "Philosophical Investigations": Generally Known as the Blue and Brown Books*. 2nd edition. New York: Harper & Brothers.

Woodward, James. 1989. "The Causal Mechanical Model of Explanation." In *Scientific Explanation*, edited by Philip Kitcher and Wesley C. Salmon, 357–83. Minnesota Studies in the Philosophy of Science, vol. 13. Minneapolis: University of Minnesota Press.

Woodward, James. 2000. "What Is a Mechanism? A Counterfactual Account." *Philosophy of Science* 69 (S3): S366–77.

Woodward, James. 2003. *Making Things Happen*. Oxford: Oxford University Press.

Woodward, James. 2011. "Mechanisms Revisited." *Synthese* 183 (3): 409–27.

Wouters, Arno. 2007. "Design Explanation: Determining the Constraints on What Can Be Alive." *Erkenntnis* 67 (1): 65–80.

Wright, Cory, and William Bechtel. 2007. "Mechanisms and Psychological Explanation." In *Philosophy of Psychology and Cognitive Science*, edited by Paul Thagard, 31–79. New York: Elsevier.

Yablo, Stephen. 1992. "Mental Causation." *Philosophical Review* 101 (2): 245–80.

Ylikoski, Petri. 2018. "Social Mechanisms." In *The Routledge Handbook of Mechanisms and Mechanical Philosophy*, edited by Stuart Glennan and Phyllis Illari, 401–12. London: Routledge.

Zednick, Carlos. 2011. "The Nature of Dynamical Explanation." *Philosophy of Science* 78 (2): 238–63.

Zurek, Wojciech Hubert. 2003. "Decoherence, Einselection, and the Quantum Origins of the Classical." *Reviews of Modern Physics* 75: 715–75. doi:10.1103/RevModPhys.75.715.

Index

Printed and bound by CPI Group (UK) Ltd, Croydon, CR0 4YY